MILITARY TRAINING AIRCRAFT OF TODAY

MILITARY TRAINING AIRCRAFT OF TODAY

ROY BRAYBROOK

Foulis

Haynes

Preceding pages: Embraer Tucanos of the Venezuelan Air Force, illustrating the silver paint scheme used in flying training and the camouflage of close support. (Embraer)

ISBN 0 85429 598 4

A **FOULIS** Aviation Book

First published 1987

Published by:
Haynes Publishing Group
Sparkford, Nr. Yeovil, Somerset
BA22 7JJ, England.

Haynes Publications Inc.
861 Lawrence Drive, Newbury Park,
California 91320, USA.

British Library Cataloguing in Publication data
Braybrook, Roy
    Today's military training aircraft : a
    comprehensive international survey.
    1. Airplanes, Military
    I. Title
    623.74.'6   TL685.3
    ISBN 0-85429-598-4

Library of Congress catalog card number
    87-81698

Editor: Mansur Darlington
Page Design: Mike King
Printed in England by: J.H. Haynes & Co. Ltd

Contents

Chapter 1

An Introduction to Pilot Training

MILITARY PILOTS are trained to fly aircraft in one of three principal aircraft categories. The first, and in many respects the most demanding, is what Britain's RAF terms the 'fast jets', ie, fighters such as the F-15 and Tornado F2, attack aircraft such as the AV-8B and Buccaneer, and reconnaissance aircraft such as the RF-4E and Jaguar. The second category is large multi-engined subsonic aircraft, ie, transports such as the C-130, tankers such as the KC-135, and maritime patrol aircraft such as the Nimrod and P-3C. Where applicable, this category also includes subsonic bombers such as the B-52 and Tu-16. The third category is rotary-wing aircraft.

In some air forces all student pilots are trained as though they were to graduate to the fast jets. At the end of their flying training those graduates considered the most suitable for fighters, attack aircraft and high-speed reconnaissance aircraft are 'creamed off', and the remainder are retrained either for multi-engined aircraft or helicopters. This approach, which has traditionally been supported by the USAF, is good for the students' morale and ensures the highest possible standards in the fast jet graduates, but it is very expensive. Other services consequently prefer to divide the students into three specialist streams as soon as they have completed basic flying training on fixed-wing aircraft. Those students chosen for the multi-engine and helicopter streams then complete their training on light transports and small rotary-wing aircraft.

The equipment used to train these latter streams is illustrated by the RAF's Jetstream and Gazelle/Wessex combination. In essence, what is required is a dual-control aircraft that is inexpensive to operate, and will

British Aerospace Hawks in a variety of RAF colours. In the foreground, an advanced flying trainer from No4 FTS. In the middle, an aircraft from No1 Tactical Weapons Unit at RAF Brawdy, with the markings of No79 Sqn. In the rear an aircraft in the medium grey paint scheme developed for the Hawk's war-role of low level airfield defence (British Aerospace).

Generations of RAF trainers. The Hawk in the right foreground has replaced the Gnat on the right in advanced flying training and the Hunter T7 on the left in weapons instruction. The Jet Provost at centre rear is to be replaced in the basic training role by the Shorts Tucano. (British Aerospace)

prepare the student for the role of second pilot on an operational type such as a C-130 or Chinook. The point is that the aircraft used to provide dedicated training for students who will (after final graduation) fly multi-engined aircraft and helicopters were not designed specifically as trainers. They are therefore not discussed in this book, which is restricted to those aircraft that are employed by the fast jet stream, and were designed from the outset primarily for the pilot training role. Secondary operational roles, however, may also have been taken into account in their design in order to expand the potential market for the product.

The training syllabus for the fast jet stream consists of three phases of flying tuition, which may be followed by a short course to introduce the student to tactics and weapons, although many services carry out both tactical weapons instruction and operational conversion with the squadrons to which the pilots are posted on graduation.

The first phase in **primary flying training** or grading, and in WWII parlance was *'ab initio,'* since the student came to this with (supposedly) no previous flight instruction. It typically covers the first 20-25 hours of the syllabus, and is intended to establish whether the student really has the necessary aptitude to become a military pilot in a reasonably short time.

Typical of primary trainers is this Bulldog 128 of the Royal Hong Kong Auxiliary Air Force. (British Aerospace)

This phase eliminates (for example) those who get airsick, lack co-ordination, and have poor judgement of heights (and thus have great difficulty in making smooth landings). It normally takes the more successful students to the stage at which they can make a solo flight, involving a simple circuit of the airfield.

The primary training phase has a high failure rate, and is therefore normally carried out on the least expensive aircraft consistent with the grading task. A typical trainer in this category has a 200 hp piston engine, side-by-side seating, and a fixed landing gear.

For the student who passes through the grading process successfully, the next phase is 100-150 hours of **basic flying training,** which is generally carried out on a turboprop or jet aircraft. During this phase the student learns such skills as aerobatics, night flying, formation flying, and cross-country navigation. Aircraft used for basic flying training generally have maximum level speeds in the range 250-400 knots (465-740 km/hr), the lower figure being achieved by the more powerful types of turboprop aircraft and the upper by turbofan-powered trainers.

Side-by-side seating is illustrated by this cockpit photograph of the Saab 105. (I. Thuresson, Saab-Scania)

Traditionally, side-by-side seating has been favoured for this role, since close contact between student and instructor was felt to be essential, but most current basic trainers now have tandem seating. The transition to the latter arrangement has been eased by the use of simulators, allowing the student to become familiar with the cockpit before he ever leaves the ground. Tandem seating has the advantage of providing a higher aircraft performance, and of acclimatising the student to the seating arrangement of a fast jet. It also facilitates formation flying, since the student can see clearly on both sides. Nonetheless, it is interesting that both the USAF and RAAF have stated a preference for side-by-side seating in a basic trainer.

The third phase of the fast jet student's syllabus is **advanced flying training,** which usually lasts 100-150 hours. In essence, the course repeats all the exercises that were included in the basic flying phase, but at speeds more appropriate to a modern combat aircraft, and (in Western Europe) with the emphasis on low level operation. Tandem seating is the general rule for this phase, although the two-seat version of the Hunter has side-by-side seating. The leader among modern advanced trainers was

One example of tandem-seat basic jet trainers is the Czech Aero L-29 Delfin. (Roy Braybrook)

the T-38, which can reach Mach 1.30 in level flight at altitude. The use of afterburners is expensive, however, and (more significantly) reduces flight endurance to a very short time. In consequence, reports indicate that USAF pilot training now includes only one supersonic sortie on the T-38.

The RAF originally planned a supersonic advanced trainer similar to the T-38, which (combined with a French Air Force requirement for a STOL ground attack aircraft) resulted in the comparatively expensive Jaguar. However, most air forces (including the RAF) have now decided to accept subsonic advanced trainers to economise on operating costs. This accounts for the Saab 105, and the Hawk, MB. 339 and Alpha Jet. The only real exception to this rule has been Japan's Mitsubishi T-2, which must be the most expensive training aircraft in the world.

Although operating costs can be said to dictate the use of subsonic trainers, their employment is made more acceptable by the current NATO emphasis on low level operation. In this context present-day combat aircraft may be able to reach speeds of Mach 1.1-1.2, but only transiently. They cruise at low level at speeds in the region of 540-600 knots (1000-1110 km/hr) and may well attack ground targets somewhat slower in order to allow for last-minute course corrections. A trainer such as the Hawk has a maximum low level speed of around 560 knots (1040 km/hr), and thus can cruise almost as fast as a modern operational aircraft.

Prior to the Hawk, the RAF used the diminutive Gnat for advanced flying training, although it could not accommodate taller pilots or be employed for weapons instruction. (MoD)

Tactical weapons training teaches the student how to plan and perform the basic air-air and air-ground missions. It includes air combat manoeuvres and air-air gunnery against towed targets, using either sleeves on which the bullet pattern is recorded by holes (coloured by painting the projectiles prior to loading in the aircraft) or smaller targets carrying acoustic miss-distance recorders. Air-ground attacks normally include gunnery, firing unguided rocket projectiles (RPs), and dropping practice bombs designed to simulate the ballistics of free-fall or retarded full-scale weapons.

Tactical weapons instruction is often carried out on the same aircraft type as is used for advanced flying training (eg, Hawk, Alpha Jet). One exception was the Gnat Trainer, which was completely unsuited to weapons training, for which the RAF employed the Hunter two-seater. The latter also had the advantage of side-by-side seating, which gave both student and instructor an excellent view over the nose. The Gnat and Hunter were later replaced by the Hawk, which is far more economical, but flies weapon training missions 50 knots (93 km/hr) slower than the Hunter and, having tandem seating, is somewhat limited in terms of the instructor's field of view over the nose.

The Hawker Hunter T7, though inferior to the Indian Mk66 in performance and armament, was one of the best weapons training aircraft ever produced. (Roy Braybrook)

13

Demands for Cost-Reduction

If the slice of the defence budget associated with pilot training is divided by the number of graduates that appear each year, then the resulting average cost per successful student amounts to more than half the price of a modern advanced trainer. In early 1987 a figure of £3.7 million was given in a Parliamentary reply on the cost of training a Sea Harrier pilot from scratch; at the same time, an export Hawk had a unit flyaway price in the region of £6.0 million.

It follows that the overall cost of pilot training is a major item of defence expenditure, and one that has a serious impact on what is generally regarded as the most important slice of the budget, ie, the amount spent on new combat aircraft. There is consequently world-wide interest in reducing pilot training costs.

There appears to be widespread agreement on how the cost of flying training may be minimised, although in practice, training schemes differ widely, due to variations in emphasis, the need to continue with existing equipment to the end of its economical life, and numerous local factors.

The employment of advanced trainers in operational roles is illustrated by these two Hawk Mk60s from the Air Force of Zimbabwe, armed for ground attack and low level air defence. (British Aerospace)

The basic rules of cost-reduction are:

1) to reduce the number of different trainer types in the syllabus to a rational minimum, which is probably three.

2) to increase the use of flight simulators, which have only a fraction of the hourly operating cost of the actual aircraft.

3) to make the maximum use of low-cost trainers in the basic category, in order to reduce the time spent on the more expensive advanced trainers.

4) to minimise the cost of students who fail the course by identifying likely failures at the earliest possible stage.

5) to 'stream' the students at an early stage, in order to minimise the flying hours on advanced trainers.

6) to establish a syllabus based on a rational combination of trainer types, so that the student is taken through the course in a series of manageable steps, with the changeover points biased in favour of reducing the time spent on the more expensive aircraft.

7) to use the advanced trainer for some operational roles, such as close support, tactical reconnaissance, low level airfield defence, and anti-ship strikes. Such commonality reduces the cost of training pilots and groundcrew, and the funds tied up in spares holdings.

There have been several unsuccessful attempts to reduce pilot training costs. For example, various air forces have experimented with the use of low-cost civil trainers for the primary phase, apparently in the belief that there is little difference between piston-engined civil and military training

aircraft. In reality there is a divergence between them, with civil aircraft tending toward engines in the region of 70-80 hp, while military aircraft are moving toward 300 hp or more.

What has been found when the services made trials with civil trainers is that such aircraft are too easy to fly. Any student anxious to succeed, naturally pays for a few hours of tuition at his local flying club before entering the service, and consequently has no difficulty in impressing his instructor with what appears to be a natural talent for flying. The result is that likely failures who should have been spotted in this first phase pass through with flying colours, only to be washed out later, when far more money has been wasted. Whereas civil trainers can emphasise minimum operating cost, the role of a military primary trainer is to show up any fundamental shortcomings in the student, and this requires higher performance and special handling characteristics. One of the best trainers ever built was the T-6 Havard, which was very difficult to fly really well, although no-one would suggest a return to a tailwheel undercarriage to present students with a greater challenge.

A number of air forces have also experimented with single-type training, in order to eliminate completely the time spent in converting from one aircraft to another. This might be regarded as an over-reaction against some early post-war flying training schemes, in which six or seven different aircraft types were used. For most air forces, however, single-type tuition is probably far more expensive than a syllabus involving three different types for the three phases of training.

Single-type syllabuses are undoubtedly possible, assuming a reasonably high quality of student and an aircraft with a wide speed range, eg, 100-500 knots (185-925 km/hr). The basic problem is that using a single trainer type is financially attractive only to very small air forces (since three MB.339s might be cheaper to operate than one Epsilon, one Shorts Tucano and one MB.339), yet the really small air forces generally have students who are less well prepared for flying training. Such students are often badly educated, have never driven a car, and quite possibly have never flown in any type of aircraft. They are thus in no sense prepared for an aircraft that falls out of the sky at around 100 knots (185 km/hr).

Ironically, it seems to be the larger air forces that are more interested in reducing the number of different trainer types to an absolute minimum, despite the fact that they operate fleets of a quite economical size. For example, the Soviet Air Force is known to have experimented with all-through training on the Czech-built L-39, taking students from the *ab initio* stage to MiG-21 conversion, and may now be using this one-type syllabus for a significant proportion of its student pilots. In contrast the much smaller Czech Air Force still uses the old L-29 for basic jet training, although in the longer term (as the L-29s are retired) the L-39 may carry out both basic and advanced training, following grading on a propeller-driven primary trainer.

Even if the Soviets train all their fast-jet pilots on an all-through L-39 syllabus, this will still not prove that single-type training is a good idea.

The L-39 may well sort out the likely failures very quickly, but in real terms it is far more expensive to operate than a conventional primary trainer. In addition, it seems highly likely that the Soviet Air Force buys its trainers from Czechoslovakia at a much lower price than even the incredibly low $1.6-2.0 million charged in the overseas market, hence the economic optimum would not apply elsewhere. Thirdly, it must be borne in mind that a lot of advanced flying training is actually carried out on the MiG-21U at the operational squadrons, hence the Soviet Air Force would really be using two-type training.

Chapter 2

Recent Trainer Developments

THE POST-WAR history of military trainers may be regarded as alternating waves of basic and advanced training aircraft. Initially there was a wave of advanced trainers such as the two-seat Meteor, Vampire and T-33, all based on first generation jet fighters. Then it was decided that students should graduate to jets at an earlier stage. Fortunately, a whole generation of small, lightweight gas turbines was coming along, some of which were being developed for missile and drone applications. Responding to the market demand, and adopting these new small engines, the manufacturers produced in the 1950s a whole series of basic jet trainers. The highlights of this generation included the CM-170 Magister, which first flew on 23 July 1952, the Jet Provost that followed on 26 June 1954, and the T-37 that followed on 12 October of that year. Italy's MB.326 flew on 10 December 1957, the T-2 (Buckeye) on 31 January 1958, and the Czech L-29 on 5 April 1959.

These basic jet trainers proved to be highly successful: many are still in service over 30 years after their prototypes left the ground. Attention next turned to advanced trainers, which offered cost savings in comparison with two-seater derivatives of fighter aircraft.

It is debatable which aircraft represented the first of the new advanced trainer generation. The **Saab 105** first flew on 29 June 1963 with two diminutive Turboméca Aubisque turbofans, and entered service two years later as the SK 60 with the Swedish Air Force. More powerful General Electric J85 turbo-jets were installed in a trials aircraft in 1967, and at this stage the Saab 105 became a very attractive proposition. Swedish restrictions on arms sales, however, limited exports to a single batch for

One of the most successful of the first generation of advanced jet trainers was the Lockheed T-33, photographed here in JASDF markings at Hamamatsu airbase. (Roy Braybrook)

The US Navy's standard basic jet trainer is the North American (now Rockwell) T-2 Buckeye. (BAe)

Japan developed its own basic jet trainer, the Fuji T-1. Ironically, students progressed from the swept-wing T-1 to the straight-wing T-33. (Roy Braybrook)

The Saab 105, shown here in the form of Swedish Air Force Sk60s, is an advanced trainer that never achieved the export success it deserved. (A. Andersson, Saab-Scania)

Austria, the final delivery taking place in 1972. There is no question that the Saab 105 with J85 engines deserved to achieve a far greater commercial success.

Disregarding the Saab 105, the Czechs claim that their **Aero L-39 Albatross** represents the first of the new generation of trainers, having made its first flight on 4 November 1968. It is questionable whether the L-39 is really fast enough for the advanced training role, but it certainly represented a new departure in the sense that it made use of a low-cost turbofan engine (the Ivchenko AI-25TL developed for the Yak-40 feederliner), and thus combined excellent fuel economy with low R&D costs and low-risk development. At the time some observers doubted whether commercial powerplants would prove suitable for aerobatic military applications, but no serious problems appear to have arisen. This may have been due to the fact that the engines adopted are comparatively short, rigid units that exhibit little distortion under G-loads.

If the L-39 is regarded as a basic trainer, then the first of the new advanced trainers was perhaps the Franco-German Dassault-Breguet/ Dornier **Alpha Jet**, which first flew on 26 October 1973. Combining swept wings that permitted dive speeds in excess of Mach 0.90, and moderate bypass engines developed specifically for this aircraft, the Alpha Jet offered economical advanced flying training and a useful secondary close support capability (as specified by the German Air Force).

The British Aerospace **Hawk**, which followed almost a year later, making its first flight on 21 August 1974, might be summarised as a single-engined, low-wing equivalent of the twin-engined, high-wing Alpha Jet, but it did provide improvements in certain areas. The single engine and generally simpler systems led to somewhat lower costs, and the fact that this engine (the Adour) had considerable background experience in the Jaguar made it highly reliable from the outset. The Hawk was designed to a much more severe fatigue spectrum of loads than any of its competitors, which again promised to reduce operating costs. In addition, the Hawk was designed from the outset not only for flying training and ground attack duties, but also for weapons instruction. There were consequently provisions for gunsights in both cockpits, whereas a gunsight in the rear cockpit of an Alpha Jet is a difficult installation (due to the actuator for the front hood being on the aircraft centreline).

In the late 1960s it was widely estimated that there was a market for around 5,000 advanced trainers and light attack aircraft. The manufacturers of the Alpha Jet and Hawk therefore expected to sell about 1,000 units of either type, and other constructors were encouraged to enter what appeared to be a very lucrative field. Of the many other types that appeared, three that certainly warrant mention are Spain's C-101, Italy's S.211, and Argentina's IA-63. The CASA **C-101** has a Garrett TFE731 turbofan, and first flew on 29 June 1977. The SIAI-Marchetti **S.211** has a Pratt & Whitney Canada JT15D (likewise developed for business jets) and first flew on 10 April 1981. It was probably the first trainer to make use of a supercritical wing section. The FMA **IA-63** first flew on 6 October 1984, and

Below left: The L-39 Albatross is regarded by the Czechs as the first of the turbofan trainer generation. (Roy Braybrook)

has a TFE731 similar to that of the C-101. However, being a much smaller aircraft, it has a somewhat higher performance.

With the notable exception of the L-39, the new jet trainers have turned out to be very expensive aircraft to buy and operate. This probably accounts for the fact that export sales have been very disappointing: none of the Western products shows any indication of approaching the 1,000-aircraft production run that was originally hoped for. The most successful is the Hawk, exports of which are now expected to exceed 500 units, but only if 307 McDonnell Douglas T-45s for the US Navy are included. In reality, sales to America were not envisaged in the initial market estimates, and the British contribution to the T-45 airframe is less than 20 per cent.

Part of this general failure to achieve large-scale production runs is undoubtedly due to NATO's lack of an effective policy on trainer equipment commonality. In contrast, most Warsaw Pact countries have agreed to accept the Czech L-39 as their standard jet trainer (Poland being the principal exception), resulting in what has amounted to a domestic market in excess of 1,000 units. Total production of the L-39 is believed to have exceeded 1,500 aircraft.

The Hawk is seen here in the colours of No4 Flying Training School at RAF Valley in North Wales. (Geoffrey Lee, British Aerospace).

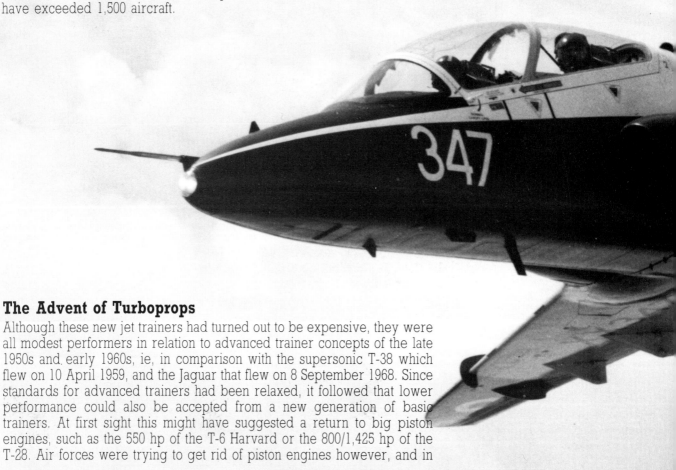

The Advent of Turboprops

Although these new jet trainers had turned out to be expensive, they were all modest performers in relation to advanced trainer concepts of the late 1950s and early 1960s, ie, in comparison with the supersonic T-38 which flew on 10 April 1959, and the Jaguar that flew on 8 September 1968. Since standards for advanced trainers had been relaxed, it followed that lower performance could also be accepted from a new generation of basic trainers. At first sight this might have suggested a return to big piston engines, such as the 550 hp of the T-6 Harvard or the 800/1,425 hp of the T-28. Air forces were trying to get rid of piston engines however, and in

The PC-7 first flew in 1966, blazing the trail for the turboprops that followed a decade later. (Pilatus)

some parts of the world AVGAS was simply not available. On the other hand, the Pratt & Whitney Canada (P&WC) PT6A turboprop was in widespread use and was establishing an excellent reputation. A handful of people recognised an opportunity to create a new class of relatively low-cost basic trainers, that would be well-matched to the new advanced trainers that were appearing.

Although the new turboprop trainers 'took off' only in the late 1970s, the foundations for this generation were laid in the previous decade, when Switzerland's Pilatus designed the **PC-7** as a derivative of the piston-engined P-3. The prototype PC-7 had its maiden flight on 12 April 1966, with a modified PT6A-20 engine, but the first production PC-7 did not fly until 18 August 1978, and this Turbo-Trainer (a name that did not catch on) did not enter service until the following year.

In the meantime the PC-7 had been overtaken by the Beech **T-34C Turbine Mentor**, which first flew on 21 September 1973, with deliveries for the US Navy following three years later. The T-34C is powered by the P&WC PT6A-25, the first fully aerobatic production engine of the series. In the US Navy T-34C it is torque-limited to 400 shp to ensure a long life and low operating costs, but in the export T-34C-1, which is stressed for external stores, this limit is raised to 550 shp. The basic engine is capable of 680 shp, hence these nominal ratings can be maintained to very useful altitudes, with consequent benefit to climb performance. The PT6A-25A of the PC-7 is also rated at 550 shp, and is somewhat lighter, as the result of the use of some magnesium castings. Both the PC-7 and T-34C-1 are capable of speeds marginally in excess of 200 knots (370 km/hr) in level flight.

In the late 1970s these two turboprops shared the export market between them. The PC-7 was slightly more expensive due to Swiss production costs, but Pilatus argued with conviction that the cost difference was more than offset by the PC-7's vastly superior handling characteristics. Beech had developed the T-34C under US Navy funding and to a tight deadline, whereas Pilatus had had more than 12 years to perfect the PC-7.

Both the PC-7 and the T-34C sold reasonably well, but it was clear from the outset that there were dangers associated with the whole concept of what may be termed derivative aircraft. The 550 shp PC-7 was derived from the 260 hp piston-engined P-3. The 400/500 shp T-34C was derived from the 225 hp piston-engined T-34B. Since the PT6 series, which first ran on the bench in 1959 and first flew in 1961, was potentially capable of more than 1,000 shp, it was obvious that someone starting from scratch could develop a trainer of significantly higher performance. Such an aircraft however, would be much more expensive to develop, and there was some doubt as to whether a market existed for a high-priced turboprop trainer.

On 16 August 1980 Brazil's Embraer **EMB-312 Tucano** had its maiden flight, powered by a 750 shp PT6A-25C. The EMB-312 produces a maximum level speed of 242 knots (448 km/hr) and it is a much larger aircraft than the PC-7, with a sufficiently spacious cockpit for ejection seats, the rear one of which is raised to improve the instructor's field of view. For most air forces the EMB-312 may have provided more power than was required for flying training, but for those air forces that needed a close support or COIN aircraft, it represented a more effective choice than the PC-7. Its greater power allowed it to lift heavier loads out of a short airstrip, and its ejection seats gave the pilot a better chance of survival in the case of defensive fire.

The other innovation in the turboprop trainer field was the attempt by various manufacturers to extend the turbine spectrum to somewhat lower powers, using the Allison 250 series. The first such aircraft was probably the Rheinflugzeugbau **Fantrainer,** which first flew on 27 October 1977, and was unusual in having a centrally-mounted engine driving a ducted pusher fan. The first conventional trainer with an Allison 250 turboprop was the SIAI-Marchetti **SF.260TP,** which flew on 1 July 1980. This was unusual only in the sense that the aircraft was unchanged from the 260 hp piston-engined original aft of the firewall. This was convenient for the manufacturer, but it also left the SF.260TP somewhat short of fuel, and this may account for the aircraft's limited success. The Allison engine was derated to 350 shp to minimise the need for changes to the aircraft. Other trainers followed suit, including India's HAL **HTT-34**, Chile's ENAER **Aucan**, Japan's Fuji **KM-2D** and Finland's Valmet **L-90TP**.

The various trainers powered by Allison 250 turboprops have so far had little impact on the market, although some piston-engined SF.260s are being re-engined in the field, a conversion that is particularly simple in this case. It is also worth noting that other engine manufacturers are aiming at the lower end of the turboprop trainer range, the 450 shp Turboméca TP319 being tested in the first Epsilon prototype, and the Teledyne Continental TP500 reportedly being developed for ratings in the 250-520 shp range.

On the other hand, engine ratings in the region of 300 hp still represent a grey area between pistons and turboprops. The old-fashioned piston engine may be unpopular in terms of its demands for AVGAS, but it is basically a less expensive power-plant than a turbine engine, which needs special materials to withstand very high temperatures. In the piston-engined category, recent years have seen the advent of France's 300 hp Aérospatiale **Epsilon** and Chile's similarly-powered ENAER **Pillán**, Romania's 290 hp ICA **IAR-831**, (following the 750 shp IAR-825TP turboprop), Poland's 325 hp **PZL-130 Orlik**, and the announcement of Yugoslavia's 295 hp UTVA **Lasta**. Reciprocating engines are thus far from finished in the 300 hp category.

To revert to the mainstream of turboprop trainer developments, the advent of the Allison 250-engined trainer series had little impact on exports of the PC-7 (which was far more dependent on overseas sales than the T-34C), but the EMB-312 Tucano represented a serious threat in a market that was already showing signs of approaching its limit. At this point Pilatus might very well have embarked on a further development of the PC-7 to accommodate the 750 shp PT6A-25C and ejection seats of the Brazilian aircraft, in order to retain a competitive position. Instead, the

Swiss company decided on a derivative design of the PC-7, using an even more powerful variant of the PT6A, to leapfrog the EMB-312 Tucano and attack a completely new sector of the trainer market, competing directly with the lower end of the turbofan range.

The resulting **PC-9** was based on the PT6A-62, derated from 1,150 shp to 950 shp, and a redesigned cockpit that would accommodate two ejection seats. Although virtually a complete redesign, the PC-9 is still a relatively light aircraft, and the vast injection of power produces a scintillating performance. Sea level rate of climb is 4,000 ft/min (20.3 m/sec), and maximum level speed increases from 268 knots (497 km/hr) 'on the deck' to 300 knots (555 km/hr) at 20,000 ft (6,100 m). The PC-9 first flew on 7 May 1984.

At first sight Pilatus had created a new trainer category, and could reasonably have expected to have enjoyed a monopoly of that market sector for a few years, until the EMB-312 Tucano was developed to compete. In the event no such monopoly eventuated, because the RAF issued a requirement for a new turboprop trainer to replace its Jet Provost. In typical Whitehall fashion, this called for an improvement over the fastest turboprop then flying, ie, the EMB-312 Tucano. The issuing of this air staff target (AST.412) in June 1983 thus encouraged any manufacturer to produce a turboprop trainer that was directly competitive with the forthcoming PC-9. The AST called for a 240 knot (445 km/hr) aircraft with a 12,000 hour safe fatigue life to the Jet Provost spectrum, and a stepped tandem-seat cockpit similar to that of the Hawk, to which successful students would progress. It was also emphasised initially that the

The P.164-109 was British Aerospace's turboprop submission to AST.412. It was later dropped in favour of BAe supporting the Pilatus PC-9. (BAe, Brough)

successful contender would be an 'off-the-shelf' buy, in order to ensure that it would be in service when the Jet Provost reached its retirement date in 1989. 'Assisted escape' was called for, since the British countryside was felt to be no longer suitable for forced landings.

Success in the AST.412 contest promised an order for 130 aircraft, together with the very worthwhile endorsement of the RAF for what might be described as the world's finest basic trainer. The result was submissions on 17 different aircraft types from 15 companies, presented in November 1983. On 13 March 1984 the *Financial Times* published what appeared to be a leak from Whitehall, setting out a shortlist consisting of Britain's Firecracker, Brazil's EMB-312 Tucano, Switzerland's PC-7 and Australia's AA-10. This newspaper report also stated that the final decision would be made in the summer. Four days later the shortlist was confirmed in Parliament with the slight refinement that (as might have been guessed) the last two contenders were the more powerful PC-9 rather than the PC-7, and the tandem-seat A-20, rather than the A-10, which had side-by-side seating.

In view of the need for an offset programme, a series of partnerships were then announced, with British Aerospace (BAe) supporting Pilatus on the PC-9, Shorts supporting Embraer, Hunting Engineering (the original builder of the Jet Provost) supporting the Firecracker, and Westland supporting the Australian Aircraft Consortium. The four shortlisted teams were invited to bid against Specification T301 D&P in June 1984. There were some complaints about the way the contest had been organised, notably from manufacturers of turbofan trainers (the S.211 and T-46), who had been encouraged to respond to the AST, despite the fact that the RAF evidently had no intention to buy such an aircraft. It had widely been anticipated that two turbofans and two turboprops would be shortlisted, but in fact all turbofans were eliminated at this stage.

The shortlist was further reduced on 18 November 1984, when it was formally announced that the Firecracker and A-20 had been eliminated on cost and performance grounds. However, further submissions were then made by Hunting Firecracker Aircraft and the AAC-Westland consortium, and 'best and final' offers on all four trainers were invited for 31 January 1985.

What should have been a perfectly straightforward procurement contest for a comparatively small contract (by the standards of any major NATO power) ended with bitter complaints that MoD had 'moved the goalposts' and that the politicians who had made the final decision had used trickery to justify their choice.

The main complaint regarding modifications to the AST was that it had originally stated that a speed of 210 knots (389 km/hr) was essential and 240 knots (445 km/hr) desirable, but that this was changed to a flat demand for the latter figure, even when allowances were made for the normal loss of performance during the life of the aircraft. This modification effectively ruled out any proposal that lacked a 1,000 shp engine. In particular, it knocked out the Firecracker, which had been uprated from 550 to 750 shp to achieve 233 knots (432 km/hr). It also ruled out the A-20, which had been uprated from 750 to 850 shp to achieve 238 knots (441 km/hr).

In effect, the AST began as a modest improvement on the EMB-312 Tucano and gradually changed to demand the performance of the PC-9.

Shorts, bidding as Embraer's partner on variants of the EMB-312, responded with a series of more powerful engines. Moving up from the 750 shp PT6A-25C to the 850 shp – 25C/2 gave 240 knots (445 km/hr) at sea level. As MoD upped its requirement, however, Shorts offered the choice of the 930 shp PT6A-25C/3, giving 254 knots (471 km/hr), or the 1,100 shp Garrett TPE331-12B, giving 268 knots (497 km/hr).

In making its choice, the RAF appears to have concluded that the Firecracker was too small for its needs, and that its low aspect ratio wing did not really give the 'jet-like handling' that some had claimed. The A-20 was a more attractive proposition, being derived from the A-10 (then under development for the RAAF) specifically to meet AST.412, but it was far removed from the 'off-the-shelf' aircraft desired.

The Tucano was very popular with RAF engineers for its large size, which made possible easy access to all components. The performance of the Shorts version of the Tucano met the requirement, but the handling of the aircraft needed further refinement to satisfy RAF demands, mainly in regard to stall-warning and throttle response in the approach. In the event, the Garrett-engined Tucano was not to fly until 14 February 1986 (in Brazil). The fully-modified Shorts Tucano was also to have structural strengthening, a stronger undercarriage, a windscreen between the two cockpits, an airbrake to increase the rate of descent, a cockpit revised to simulate that of the Hawk, British equipment, and a four-blade propeller. With all these modifications there was no way that the Shorts Tucano could be considered an 'off-the-shelf' aircraft as specified. Any shortcomings in the new variant could undoubtedly be ironed out, but it was impossible to be certain that this would be completed on the specified timescale. On the other hand the PC-9 was judged to be totally acceptable as tested, and it therefore became the RAF evaluation pilots' first choice, with the Shorts Tucano in second place due to reservations on the timescale on which it would be ready for service.

Despite the fact that 'best and final' offers had been submitted by British Aerospace and Shorts (on the PC-9 and Tucano respectively) on 31 January 1985, the bidders were invited by MoD to revise their offers. According to MoD, both were informed on 13 March that these revisions had to be made by mid-day on the 14th, though BAe subsequently denied receiving this information. On the morning of the 14th, Shorts offered to supply 130 modified Tucanos for £125 million. On the 19th BAe offered to supply 130 PC-9s for £119.5 million. On the 21st the then Defence Minister Michael Heseltine announced in Parliament that the Shorts bid had been accepted as the lowest cost tender received prior to the deadline.

This decision was extremely embarassing for BAe, which company was in the middle of a campaign to sell a far more valuable batch of Hawks to Switzerland. On the other hand, if the contestants were viewed from purely a political aspect, it had been clear from the outset that all the principal factors favoured the Shorts Tucano bid. The most important consideration was almost certainly to reduce unemployment in Northern Ireland, in addition to which HMG wished to prepare Shorts for

privatisation. Thirdly, Britain owed a considerable debt to Brazil for that country's help at the time of the Falklands crisis of 1982.

On the other hand, marketing considerations appeared to favour the PC-9. Aside from BAe's need for support for the Hawk in the Swiss context, it was argued that the Tucano offered only limited prospects of follow-on sales, since most of the potential market was already covered by production lines in Brazil and Egypt, which could probably produce aircraft much more cheaply than Shorts.

It may be added that many observers were surprised when on 14 January 1987 the Swiss Federal Military Department announced its decision in favour of the Hawk, rather than the Alpha Jet. The factors cited in favour of the Hawk were 'its clearly lower price, its strong airframe designed for a long service life, and its cockpit providing better visibility'. Although BAe had failed to sell the PC-9 to the RAF, the company had succeeded in including 30 PC-9s with a package of Tornadoes and Hawks sold to Saudi Arabia, and had bought 11 FFA AS.202 Bravos for its flying training school at Prestwick. The value of 20 Hawks for Switzerland was Sw Fr 395 million (around £170 million), bearing out the BAe argument that this was a much larger contract than the RAF purchase of 130 basic trainers.

The expansion by Pilatus into the 1000 shp turboprop trainer category and the effect of the RAF contest in focusing attention on this area have changed the nature of the military trainer market. Turboprops now cover a wide power range from around 350 to 1100 shp, they burn fuel that is universally available, and (unlike piston engines) they provide single-lever operation. In choosing to purchase a turboprop to replace the Jet Provost, the RAF reckoned that a turboprop trainer represented only half

The winner of the AST.412 contest was the Shorts Tucano with 1100 shp Garrett engine. (Shorts)

the flyway price of a turbofan, and only 75 per cent of the latter's hourly operating cost.

The RAF estimates, however, were almost certainly based on less powerful turboprops such as the 750 shp EMB-312 Tucano. Both the PC-9 and Shorts Tucano are significantly more expensive aircraft. For example, in 1968 a PC-9 cost $1.8 – 2.0 million compared to $1.1 – 1.2 million for the PC-7, an increase in the region of 65 per cent. It is also noteworthy that, in originally advocating a turbofan trainer to replace the Jet Provost, BAe argued that to provide equal performance a turboprop would be just as expensive as a turbofan.

At time of writing the 1000 shp turboprop basic trainers are still very new, but there already appears to be some feeling in the market place that this category has now been pushed to its limit, and that it is going to operate at airspeeds at which a pure jet would provide better handling characteristics. For example, during the RAF evaluation it was found that all the turboprops exhibited some degree of snaking at high speeds.

New Jet-Powered Basic Trainers

In the same way that the PC-9 was developed specifically to compete with turbofan traners, there is now a situation in which there is scope for a low-cost turbofan trainer developed specifically to compete with the PC-9 and Shorts Tucano. Assuming that such an aircraft can be competitive in flyaway price and hourly operating cost, then there is a great deal of force in the old argument that student pilots intended for the fast jets should be acquainted with jet propulsion as early as possible in the syllabus.

If the market is currently short of low-cost turbofan trainers, it is because suitable engines have been slow to appear. One of the first factors that sparked off interest in the idea of a small jet trainer was a study carried out in the early 1970s by a USAF group at Wright-Patterson AFB, promoting the idea of a very low-cost trainer (VLCT). The basic concept was to provide inexpensive continuation training for pilots operating very costly military aircraft. The VLCT study led to the testing of various aircraft, including the diminutive Bede BD-5J. However, it appears to have been concluded that the value of such training was very limited, and that it was better to have the pilots fly business jets, although the companion trainer aircraft (CTA) programme for B-52 crews appears to have been put on the back burner.

One result of the VLCT study was that it encouraged Caproni to develop a small trainer, the **C-22J.** The original idea was that Williams International would certificate a derivative of the F107 cruise missile engine, giving a thrust in the region of 900 lb (400 kg). When this powerplant failed to appear, Caproni looked around for alternatives, and the most likely substitute appeared to be the Microturbo TRS-18, which the company had used to give the A-21SJ sailplane a self-launch capability. The A-21SJ had first flown in 1977 with a single TRS-18 rated at 202 lb (91.6 kg). It appeared that, when allowance was made for future thrust growth, a twin-engined basic jet trainer might be based on TRS-18s, although it

The Caproni C22J is a bold attempt to develop a lightweight jet trainer, using experience from a jet-powered sailplane. (Roy Braybrook)

would initially be underpowered. In the latest C-22J brochure the engine thrust is given as 328.5 lb (149 kg), which is a major improvement over the initial rating, though it still falls short of the original F107-engined concept. The sea level maximum speed of the C-22J is 280 knots (520 km/hr), which is marginally superior to that of the best turboprops. On the other hand its initial climb rate of 1,950ft/min (9.9 m/sec) is very modest in comparison with the 3,500 ft/min (17.8 m/sec) of the Shorts Tucano and the 4,000 ft/min (20.3 m/sec) of the PC-9.

The only other aircraft in the same class as the C-22J is Microturbo's own **Microjet 200B,** a project which has now been sold to Marmande Aeronautique, part of the Creuzet group. The wooden prototype Microjet 200 first flew on 24 June 1980, a month ahead of the Italian aircraft. The pre-production 200B of mixed metal and composite construction flew on 19 May 1983. The Microjet is unique in its seating arrangement, which might be described as modified side-by-side configuration, the right-hand seat being set back to improve the student's field of view on that side. If ordered into production, it is anticipated that the engine rating will be increased to give a take-off thrust of 405 lb (184 kg).

If other manufacturers identified a potential demand for a low-cost basic jet trainer, they evidently decided that they should wait until a more suitable engine became available. The most promising powerplant in this category came into existence with the Fairchild **T-46,** a twin-engined aircraft developed as a replacement for the USAF T-37. Specifically for this project Garrett developed a new turbofan, the TFE109 of 1,330 lb (603 kg) thrust, which also provided a suitable basis for a somewhat smaller single-engined trainer, though one that could be much larger than the C-22J and Microjet. The T-46 had its maiden flight on 15 October 1985, but production plans have been cancelled. It nonetheless appears that Garrett will make the TFE109 engine available for other applications.

The most likely customer for the TFE109 engine is the Belgian Promavia company, which in 1982 conducted a market survey, pointing to the need for a basic jet trainer that would exploit the latest engine

technology in terms of low weight, low noise, low maintenance demands and reduced fuel consumption. Compared with the ubiquitous SF.260, such an aircraft would make possible a reduction in training costs, since it would allow up to 40 hours to be eliminated from the advanced flying syllabus.

Acting on the results of this survey, and recognising the potential of the TFE109, Promavia (which is basically a marketing organisation) funded General Avia in Italy to design and construct a single-engined basic trainer with side-by-side seating. The resulting F1300 **Jet Squalus** first flew on 30 April 1987. Its maximum level speed is 300 knots (556 km/hr), which is significantly faster than that for the PC-9 or Shorts Tucano. Maximum climb rate is given as 3,200 ft/min (16.26 m/set), which is somewhat inferior to the turboprop figures, though the difference may be eliminated by further development of the TFE109 or a switch to the Williams FJ44. If Promavia is correct in stating that the Jet Squalus can be sold for $1.65-2.10 million, depending on equipment fit, then it will certainly be competitive with the upper end of the turboprop range.

As will be seen from the above discussion of recent developments, military trainers are in a constant state of flux as new demands emerge and as new types of powerplant become available. What does not change is the pressure to reduce training costs.

Above left: The Microjet 200 was initially developed by the engine manufacturer Microturbo to demonstrate the usefulness of the TRS-18 engine. (Roy Braybrook)

Above: The Jet Squalus employs the latest Garrett turbofan to make possible a low-cost basic jet trainer to compete with the PC-9 and Shorts Tucano turboprops. (Promavia)

Trago Mills SAH 1 at Farnborough 1986. (Roy Braybrook)

FFA AS.202/18A at Farnborough 1986. (Roy Braybrook)

Slingsby Aviation T67M Firefly, Farnborough 1986. (Roy Braybrook)

Valmet L-70 Miltrainer at Paris 1983. (Roy Braybrook)

Aerospatiale Epsilon at Farnborough 1986. (Roy Braybrook)

FUS Flamingo with Porsche PFM 3200 engine, at ILA-86, Hanover. (Roy Braybrook)

Opposite: Poland's PZL-130 Orlik at Paris 1985. (Roy Braybrook)

ENAER Pillan at Paris 1983. (Roy Braybrook)

Opposite top: SF.260TP with unspecified underwing stores. (Siai-Marchetti)

Top: Valmet L-90TP prototype at Farnborough 1986.
(Roy Braybrook)

Above: Hindustan Aeronautics HTT-34 at Paris 1985.
(Roy Braybrook)

Opposite left: Rhein-Flugzeugbau Fantrainer (RFB).

Left: Pilatus PC-7 Turbo-Trainer at Farnborough 1986.
(Roy Braybrook)

Opposite top: PC-7s of the Royal Malaysian Air Force aerobatic team. (Pilatus)

Opposite bottom: Bolivian Air Force PC-7, serial no 250/FAB-467. (Pilatus)

Above: Norman Aeroplane Company NDN1T Turbo-Firecracker at Farnborough 1986. (Roy Braybrook)

Below: The first prototype Turbo-Orlik was a converted PZL-130, which crashed during a demonstration flight in Columbia. (Pezetel)

The EMB-312 Tucano in Brazilian Air Force camouflage. (Embraer)

Inset: The newest addition to the RAF training fleet is the Shorts Tucano T Mk1, which is replacing the Jet Provost in the basic flying training role. (Shorts)

Chapter 3

The Selection of Trainers

PILOT TRAINING schemes suffer from a 'patchwork-quilt effect' in the sense that most air forces replace one category of trainer at a time, rather than deciding on the optimum combination from scratch. The result is sometimes far from desirable. Student pilots of the Finnish Air Force go directly from the 200 hp L-70 Vinka (Blast) to the 540 knot (1000 km/hr) BAe Hawk, since the intermediate Fouga Magister has been phased out.

Likewise, those in the French Air Force now go from the 300 hp Epsilon to the Alpha Jet, which has virtually the same performance as the Hawk. These massive jumps in performance between one training phase and the next are obviously accommodated with some difficulty. In the case of Third World air forces such steps would be virtually unthinkable. The question therefore arises, as to how an air force should (in the ideal case of establishing a pilot training system from scratch) select a combination of training aircraft that will give reasonably low operating costs and acceptable steps in difficulty between the different stages of training.

Since pilot training consists of three fundamentally different stages, viz, grading, basic and advanced flying training, it is arguable that each air force should have three trainer types. It may also be argued that the basic trainer (ie, the middle type in performance terms) is also very effective as a vehicle for weeding out those students who lack the necessary qualities to become military pilots, hence the number of different trainer types may be reduced to two.

It seems likely that, in the case of air forces with very high entry standards, the least expensive form of pilot training involves two different aircraft types. At the opposite extreme, in the case of Third World services, the first stage of training should take place on a very simple aircraft, hence the syllabus should be based on at least three different aircraft types.

To the best of this writer's knowledge, the only attempt to analyse pilot training in mathematical terms was an Italian Air Force study in the 1970s, conducted under the direction of Gen Niccoló by an operations analysis team led by Maj-Gen Giorgeri. The object of the exercise was to compare basic trainer options in replacing the MB.326 as a stepping-stone between the SF.260 piston-engined primary trainer and the Aeritalia G.91T advanced trainer. The conclusion of the Italian Air Force study was that the MB.326 replacement should be based on the 4000 lb (1815 kg) Rolls-Royce Viper 632 turbojet, rather than turbofans such as a single R-R Adour or two SNECMA Larzacs.

This official study was later employed by Aermacchi as a marketing document in support of the MB.339. It was developed by that company's Dr Paolo Mezzanotte and was first presented by the eminent Dr Bazzocchi in Rome in 1975. The same type of analysis was later used by the Agusta group to quantify the cost-effectiveness of the S.211 and C-22J.

The findings of what became known as *'The Bazzocchi Report'* are not really controversial, since most experts would agree that the combination of two excellent trainers such as the PC-7 and MB.339 will provide low-cost tuition to the standard required by the majority of air forces.

The principal objection to this report is that it involves quite complicated mathematics, and it has not so far been presented in a way that allows service analysts to modify the equations to incorporate their own assumptions. What is needed is a much simpler approach, which those concerned with equipment procurement can understand, and which reflects the operator's own assessment of the various aircraft options.

The common-sense approach is to say that the advanced trainer must be selected to provide a smooth transition to whatever operational aircraft the service employs, and that the preceding trainer(s) should be selected to give acceptable performance steps. The only remaining difficulty is then to select a parameter that represents the performance or the skill-ceiling of the aircraft, so that different options may be compared.

In this writer's view, trainers are best characterized by the student's total flying hours at the point that he is ready to transition to the next stage. Thus a 200 hp primary trainer may be regarded as having a skill ceiling of 40 hours, at which point the student has little more to learn from this aircraft, and is ready to graduate to (for example) a 1,000 shp turboprop, which will usefully be employed for the next 120 hours, and is thus characterized by a skill ceiling of 160 hours. At this point he may graduate to a 540 knot (1000 km/hr) advanced jet trainer for a final 80 hours, this aircraft thus being given a skill ceiling of 240 hours.

Since student pilot standards vary considerably, it is taken for the purpose of this analysis that the syllabus relates to a typical West European air force. In selecting trainer aircraft combinations for Third World services, the same technique applies, although the number of flight hours

actually to be spent on each type would be increased by up to 100 per cent to allow for reduced learning rates. As indicated earlier, those responsible for equipment planning can make their own assessment of the skill ceiling appropriate for each of the trainer types on offer, hence the following values are proposed only for guidance.

The military trainer spectrum may be broken down into various categories, starting with the 40-hour primary trainer with 200 hp piston engine, side-by-side seating, and fixed undercarriage. Aside from various aircraft that are no longer in production (BAe Bulldog, Saab Supporter, Pacific Aerospace CT4), this category includes the Valmet L-70 Vinka,

Category 2: Aérospatiale Epsilon, with 300 hp engine, tandem seating and retractable undercarriage. (Roy Braybrook)

Slingsby T.67M-200, and the WSK-PZL M-20 Iskierka.

The next significant step may be regarded as the 70-hour primary-basic trainer, typified by 300 hp aircraft with tandem seating and retractable landing gear. Examples include the Aérospatiale Epsilon, ENAER T-35 Pillán, and the Norman Aeroplane Firecracker (with 300 hp Lycoming).

In order to extend the use of such aircraft to around 100 hours, some of them are being fitted with turboprop engines of the order of 420 shp, as instanced by the ENAER Turbo-Pillán or Aucan.

To be useful for 120 hours a trainer may need a turboprop of around 550 shp, representing the lower end of the PT6A range. Aircraft in this class include the Pilatus PC-7, Beech T-34C-1, and the PZL-130T Turbo-Orlik.

The EMB-312 Tucano introduced a more powerful variant of the PT6A, giving 750 shp, and it seems reasonable to credit such aircraft with an additional 20 hours, making a total of 140. The only other trainer in this category is the IAR-825TP Triumf.

The upper end of the turboprop range is represented by engines of around 1000 shp, as used in the Pilatus PC-9 and Shorts Tucano.

Category 3: RFB Fantrainer 400, with 420 shp turbine engine. (Roy Braybrook)

Category 4: Pilatus PC-7 with 550 shp turboprop. (Pilatus)

The lower end of the jet trainer range (in terms of actual sales) is represented by the various turbofan-powered aircraft that achieve level flight speeds in the region of 400 knots (740 km/hr), which may be used up to the 180 hour mark. Current examples include the CASA C-101 and Aero L-39.

The Aermacchi MB.339A has appreciably higher performance, reaching 485 knots (900 km/hr) in level flight. On this scale of usefulness, it may be employed for basic and advanced flying training up to the 210 hour mark.

At the top of the trainer range, the BAe Hawk and Dassault-Breguet/Dornier Alpha Jet can both attain speeds of around 540 knots (1000 km/hr) at sea level, and may be characterized as 240-hour aircraft.

This discussion has nominated nine principal trainer categories. There are, however, some aircraft types that are offered for the training role, but do not correspond to any of these categories. For example, the highly

successful SIAI-Marchetti SF.260 has a 260 hp piston engine, side-by-side seating, and a retractable undercarriage, hence it falls between the first and second categories, at around the 60-hour level.

Likewise, the Microjet 200B and Caproni C-22J diminutive jet trainers probably fall well below the top of the turboprop range, assuming that they are judged suitable for military pilot training. At the top of the performance range, the Northrop T-38 and Mitsubishi T-2 are both capable of supersonic speed in level flight, but this appears to be of very little real value. In this context it may be noted that the RAF has a transonic dive

Category 5: IAR-825TP Triumf, with 750 shp turboprop. (Roy Braybrook)

Category 6: Shorts Tucano with 1100 shp turboprop. (Embraer)

Category 7: Aero L-39, a typical 400 knot (740 km/hr) basic jet trainer. (Roy Braybrook)

capability in its BAe Hawk, but makes no use of this in the standard training syllabus.

If an air force planning to establish a new flying training programme can divide the available aircraft into a series of performance categories such as this, allocating to each class a characteristic skill-ceiling measured in flying hours, then it is a relatively simple matter to break the syllabus down into sensibly-sized parts. Assuming that the service operates a modern high-performance fighter, then the final stage of training will probably demand an advanced aircraft in at least the 210-hour category (eg, MB.339A). If the aim is for pilots to graduate directly to a very complex aircraft with a high cockpit workload (eg, Panavia Tornado), then it may be essential to use what is termed here a 240-hour advanced trainer (eg, BAe Hawk or Dassault-Breguet/Dornier Alpha Jet).

The basic problem is thus to divide the 210 or 240 hour syllabus into either two or three sensibly-sized parts. The simplest approach would be to make these parts equal, eg, to break the 240-hour syllabus into two equal parts, corresponding to 120 hours on the PC-7, followed by 120 hours on a Hawk. On the other hand the fact that the Hawk costs several times as much as the PC-7 suggests that the second phase of training should be reduced by employing a more advanced aircraft for the first phase.

As a first approximation it might be suggested that the two phases should be adjusted to give a 2:1 ratio, some 160 hours on the PC-9 being

Category 8: Soko G-4 Super Galeb, a trainer in broadly the same category as the MB-339. (Roy Braybrook)

followed by only 80 hours on the Hawk. To examine the likely effect of such a change in syllabus, it might be assumed that in mid-1986 prices a PC-7 cost $1.2 million, a PC-9 $2.0 million, and a Hawk $8.0 million. If it is further assumed that total hourly operating cost equates to one-thousandth part of the flyaway price, then the PC-7 costs $1,200/hr, the PC-9 $2,000/hr, and the Hawk $8,000/hr. A syllabus combining 120 hours on the PC-7 and 120 hours on the Hawk thus amounts to a total flying cost of $1,104,000, while 160 hours on the PC-9 and 80 hours on the Hawk amounts to only $960,000. This suggests a potential saving in the region of 15 per cent.

Category 9: The BAe Hawk, illustrated here by the first Mk51 for Finland, at the hand-over ceremony. (Roy Braybrook)

Considering first two-phase training, the proposal to locate the change-over point to give a 2:1 ratio between the first and second stages thus implies a combination of PC-9 or Shorts Tucano with a Hawk or Alpha Jet. If the demands for the graduation standard are less extreme, the 2:1 ratio would suggest 140 hours on the EMB-312 Tucano or IAR-825TP Triumf, followed by 70 hours on the MB.339A. Likewise, if the required standard can be achieved using a 400 knot (740 km/hr) trainer, then 120 hours on a PC-7 could be followed by 60 hours on an L-39 or C-101.

Turning to three-phase training, the idea of biasing the change-over points to reduce the time spent on the more expensive aircraft might favour (for example) 100 hours on the ENAER Aucan, followed by 80 hours on a C-101 and 60 hours on a Hawk. If the necessary standard could be reached in 210 hours, the syllabus might combine 100 hours on the Aucan with 60 hours on the PC-9 and 50 hours on the MB.339A. As a third option, 70 hours on the Epsilon might be followed by 70 hours on the EMB-312 Tucano and 40 hours on the C-101.

It must be emphasised that the various types of syllabus outlined above are only illustrative examples, and that in selecting a good combination of trainers it is important not only to reduce the time spent on the most expensive aircraft, but also to ensure that each type chosen is well-suited to the pilot training role. The fact that an aircraft is used as a trainer by one or more major air forces is no guarantee that its technical development has properly been completed and that its handling characteristics are suitable for the training task. There are some trainers on the market that are too easy to fly, and others that have major faults in their handling, faults that could not be cured within the time-scale and budgetary restrictions imposed.

To summarize, the essence of selecting a good combination of training aircraft is to plan a two- or three-phase syllabus using types that (on a common-sense basis) present the student with a series of acceptable steps, biasing the change-over points in favour of reducing the time spent on high performance aircraft. If the graduation standard demands advanced flying training on an aircraft in the class of the Hawk, the optimum lead-in aircraft (for a two-phase syllabus) is probably a turboprop such as the PC-9, though the less powerful PC-7 would be almost as good. A low-powered primary trainer such as the Bulldog is certainly not the best aircraft to combine with a Hawk. Once a particular broad category has been selected for procurement, the most important consideration is a thorough evaluation of the possible contenders by pilots and engineers.

Chapter 4

The Operational use of Trainers

THE RELATIVELY high cost of all military aircraft is encouraging the use of trainers for a variety of operational tasks, including close support, ground attack, anti-ship strike, and low level air defence. Aside from the fact that trainers are generally less expensive to operate than dedicated attack aircraft and air defence fighters, there is a worthwhile saving if the number of different types in the service is reduced. Using, for example, the Alpha Jet for both advanced flying training and close support naturally minimises the cost of operational conversion, and it also produces economies in spares holdings and groundcrew training. In wartime emergency, losses in operational units may be replaced by taking aircraft from the training programme, although this would clearly be undesirable in a conflict of any duration.

Virtually all trainers are now offered with underwing hardpoints, although for political reasons these may be associated with ferry tanks rather than ordnance. Whether aircraft in the primary/basic category have a realistic operational capability depends on the local environment. There have been instances in which 200 hp primary trainers have carried out useful ground attack sorties, but these have generally relied on surprise and the absence of effective air defences. A more realistic minimum attack aeroplane is probably a 550 shp turboprop such as the Pilatus **PC-7.** Such aircraft, although much faster than a traditional primarily trainer, still have the manoeuvrability required to locate small numbers of insurgents hiding (for instance) in scrubland, and they can carry a useful load.

The PC-7 has six underwing hardpoints, two of which have 'plumbing' for fuel tanks. These hard-points can take a maximum total load of 2295 lb

One reason behind the Swedish Air Force's adoption of the Bulldog was that it provided a good firing platform for tip-mounted Bofors wire-guided anti-tank missiles. (Swedish Air Force)

Right: The original 260 hp piston-engined Firecracker, armed with eight 80 mm Hispano SURA rockets. (Roy Braybrook)

(1040 kg), although the fact that there is only a 1765 lb (800 kg) difference between maximum take-off weights in the utility and aerobatic configurations may suggest that maximum external load is associated with some reduction in internal fuel. For comparison, the brochure for the Shorts **Tucano** refers to four underwing pylons with a capacity for four Mk81 LDGP bombs, weighing in total 1,235 lb (560 kg). Alternative loads include two 7.62 mm machine gun pods with 500 rounds each, four LM-70/7 rocket pods, and eight five-inch (127 mm) rockets, this last case giving a total external load of 1345 lb (610 kg).

Turboprop trainers undoubtedly have a worthwhile operational capability in a low threat environment, but most air forces probably now think in terms of jet aircraft for any operational role. There is considerable

experience of using basic jet trainers in the close support role. Israel has made extensive use of the Magister, and South Africa has many times employed the Aermacchi MB.326 Impala in ground attack missions. In Oman's Dhofar War of the early 1970s the Strikemaster was used to good effect in the close support role, until the insurgents introduced the SA-7 missile which made necessary higher transit and attack speeds.

The lower end of the modern jet trainer market (disregarding types not yet in service) is represented by the SIAI-Marchetti **S.211,** which in clean configuration reaches 320 knots (595 km/hr) at sea level. The S.211

The 300 hp Epsilon, armed with four Matra F2 launchers, each housing six Thomson-Brandt 68 mm rockets. (Aerospatiale)

can carry external loads on four underwing pylons, although the difference between clean gross weight and maximum take-off weight is only 990 lb (450 kg). Its attack potential is further limited by the fact that it has no centreline pylon and evidently no provisions for internal cannon. On the other hand its ferry range of 1535 nm (2845 km) indicates that the S.211 could perform usefully in the reconnaissance role. In the case of small trainers, it is important that a high wing mounting should be used to give adequate ground clearance for underwing stores, although this usually results in the fuselage being so low that a centreline station is ruled

out. Both the S.211 and the IA-63 Pampa follow this rule, although their load-carrying capacity is restricted by airframe strength.

Taking an example from further up the weight scale, the Aero **L-39** is over twice as heavy as the S.211, and has far more capability in the attack role. The Czech aircraft, which has been employed operationally in Afghanistan and possibly elsewhere, has a useful built-in armament in the form of a 23 mm GSh-23 twin-barrel cannon. This can be attached under the front fuselage, with 150 rounds in a box under the rear cockpit. Rate of fire is a remarkable 3,500 rd/min. Maximum armament weight is only 2665 lb (1208 kg), but the L-39 has a ferry range of 950 nm (1760 km) with the gun and two external tanks, indicating a useful strafing capability.

The SF.260TP, pictured at Paris with a typical array of weapons. (SIAI-Marchetti)

A BAe Strikemaster at Salalah in Oman during the Dhofar War, armed with two 250 lb (115 kg) bombs and 16 68 mm Hispano SURA rockets. (Peter Stevens)

The Aviojet **C-101** is in broadly the same class, but has six pylons instead of four, a maximum ordnance load of 4000 lb (1815 kg), and a removable gunpack that takes a single 30 mm DEFA cannon or two 12.7 mm machine guns. This aircraft also has an outstanding fuel capacity, giving a range of up to 2000 nm (3700 km) or an endurance of up to seven hours, eliminating the need to allocate underwing pylons to auxiliary tanks. The C-101DD is offered with a comprehensive nav-attack fit, including a Ferranti FIN-1100 stapdown inertial system and FD4503 head-up display (HUD), and a Marconi Doppler radar.

Several types of jet trainer in use with West European air forces have operational wartime roles. For examples, the Aermacchi **MB.339As** of the Italian Air Force would in war be employed in close air support and maritime support, piloted by instructors. The aircraft of the *Frecce Tricolori* aerobatic team would likewise be used for close support and anti-helicopter missions, duties for which the pilots train during the winter months. Reports indicate that five of the service's MB.339As are routinely used in the electronic warfare role. It may also be noted that during the Falklands conflict the MB.339s of the Argentine Navy were the only operational jet aircraft deployed to the islands during the course of hostilities, no other type used by Argentina being suitable for missions from the 4250 ft (1300 m) runway at Port Stanley. These Italian-built aircraft performed well in the anti-ship role, destroying one RN frigate, reportedly with API fire from their 30 mm DEFA cannon.

As in the case of the C-101, the MB.339 is now offered with full nav-attack equipment. The MB.339C fit includes a GEC AD.660 Doppler radar, Litton LR-80 strapdown inertial reference, GEC-620K navigation computer, Kaiser HUD, Fiar laser ranger, and ELT-156 radar warning receiver. The same fit is also available for the single-seat MB.339K, which has two 30 mm DEFA cannon in the front fuselage, thus leaving all six wing pylons free for ordnance. In a typical ground attack mission, the MB.339K has a HI-LO-HI radius of 335 nm (620 km) with four 500 lb (227 kg) bombs, the corresponding LO-LO radius being 180 nm (335 km).

There is a growing interest in the idea of using jet trainers (or derivatives of such aircraft) in the anti-ship strike role, the argument presumably being that the stand-off capability provided by modern air-surface missiles reduces the demand for very high performance in the launch aircraft. In the case of the MB.339 the weapon chosen is the Oto Melara Marte Mk2A, which gives a stand-off range of 16 nm (30 km). It is anticipated that the missile would be carried under one wing, while on the other side the aircraft would have a search radar (eg, SMA APS-707) in a slipper pod. A HI-LO-HI radius of 300 nm (555 km) is predicted.

In parallel with this work to develop an anti-ship strike capability, Aermacchi has been carrying out trials with the AGM-65 Maverick for attacks on point targets, and with the AIM-9 Sidewinder for self-defence. It may be noted that in April 1982, when Libya fired two surface-surface missiles against the small Italian island of Pantellaria, MB.339As were based on the island to provide close air cover.

The **Alpha Jet** is unusual in the sense that the requirements of the two launch customers specified both advanced flying training (for the French Air Force) and close support duties (for the German Air Force). In the latter context the aircraft is typically armed with a single 27 mm Mauser cannon in a ventral pack and up to six BL755 cluster weapons, each weighing 660 lb (300 kg). Reports indicate that in war German Alpha Jets would be operated as single-seaters with an ECM pack in the rear cockpit, and that they would be employed both in the close support and anti-helicopter roles.

For the export market Dassault-Breguet has developed the NGEA *(Nouvelle Génération Ecole/Appui)* version with a French nav-attack fit, including the SAGEM ULISS-81 inertial system, Thomson-CSF VE-110 HUD, and a TMV-630 laser ranger. The proposed Lancier model would have provisions for the Thomson-CSF Agave radar (or FLIR) in the nose, and enlarged external tanks, with a third 'wet' pylon on the centreline. Armament options would include three DEFA cannon for anti-helicopter duties, and one AM.39 Exocet missile for the anti-ship strike role. The fuel system of the Alpha Jet evidently allows the use of asymmetric external fuel, the Exocet being balanced by a 138 imp gal (625 litre) tank under the other wing. With two Matra Magic missiles for self-defence, the Lancier could then achieve a radius of up to 345 nm (640 km).

The BAe **Hawk** of the RAF originally had no wartime role, but approximately half of these aircraft have since been modified to T Mk1A standard, allowing the carriage of two AIM-9 Sidewinders in combination with the ventral 30 mm Aden cannon used in weapons training. In the event of war these Hawks would be used in the low level air defence of key airfields in the UK. Reports have indicated that a pair of Hawks would operate in conjunction with a radar-equipped Phantom. In an earlier plan the modified Hawks were also to be equipped to carry BL755 cluster weapons on the two underwing pylons, so that they could also be operated in the close support role, but such provisions were never introduced due to restrictions on funding.

Export Hawks have had full ground attack provisions from the outset, with five pylons designed for a total load of 7720 lb (3500 kg). The 100-series Hawk proposed for Venezuela in the early 1980s added a full nav-attack system, designed for commonality with that of the F-16, although alternative fits are now available.

Experience of marketing the Hawk indicated that a significant number of countries preferred a single-seater for any operational role, hence in 1984 BAe announced a go-ahead for the Hawk 200. The single-seat cockpit provides much better rear view, and made it possible to incorporate two 25 mm Aden cannon and a variety of sensors in the front fuselage. Since the Hawk (being a low wing aircraft) is much higher off the ground than the Alpha Jet, it is possible to carry a large missile such as the BAe Sea Eagle on the centreline. With two 190 Imp gal (865 litre) auxiliary tanks, the Hawk 200 can then perform an anti-ship strike sortie of up to 800 nm (1480 km) radius.

In a ground attack HI-LO-HI mission with two tanks and four 1000 lb (454 kg) bombs, a radius of 540 nm (1000 km) is predicted. The use of a radar such as the Ferranti Blue Vixen developed for the Sea Harrier mid-life update would allow the Hawk 200 to employ medium-range air-air missiles such as the BAe Sky Flash or Hughes AIM-120 AMRAAM in addition to short-range weapons.

Aircraft such as the Hawk 200 and Lancier are not inexpensive, and many services will undoubtedly prefer a less costly aircraft. For example, the C-101 is currently being developed for the anti-ship strike mission with

the BAe Sea Eagle missile, as a joint effort by CASA, ENAER and BAe. Bearing in mind the outstanding range and endurance of the C-101, such a development may well prove attractive.

The argument in favour of heavier and more expensive aircraft such as the Hawk 200 is that they can combine excellent warload-radius performance with high penetration speeds. In COIN operations in both Oman and Rhodesia it was found that transit and attack speeds in excess of 450 knots (835 km/hr) were required, once the SA-7 was introduced by the insurgents. Unless slower aircraft are given completely reliable countermeasures, there will remain a strong case in favour of top-of-the-range aircraft, whenever targets with modern defences are to be attacked with short-range weapons.

Left: The single-seat MB.339K, armed with two 30 mm DEFA cannon and six 500 lb (227 kg) bombs. (Aermacchi)

Even before the Hawk and Alpha Jet combined advanced flying training with close support capability, the Saab 105 served as the Sk60C in the Swedish Air Force, performing training and operational roles. The Sk60Cs shown here have panoramic nose cameras and 135 mm Bofors rockets. (I. Thuresson, Saab-Scania).

Chapter 5

Piston-engined Trainers

IN THE DECADES before the turbine engine had been exploited to anything like its present level, piston-engined trainers covered a very wide spectrum, from the 130 hp de Havilland Tiger Moth and 140 hp Stampe SV.4 to the 550 hp North American T-6 Harvard, the 730 hp Yak-11, and the 800/1425 hp North American T-28A/B.

The early post-war years saw the development of other widely-used aircraft, such as the Cessna T-41 derivative of the civil Model 172F, and the de Havilland Chipmunk, both of 145 hp. Primary trainers provided useful experience for several European industries, as illustrated by the 180 hp Saab-91A Safir, and the 190 hp Fokker S.11 Instructor. The 225 hp Beech T-34A/B was employed both by the USAF and by the USN.

In the 1950s and 60s there was considerable agreement that the primary training phase demanded an aircraft of around 200 hp, with fixed gear and side-by-side seating. Logically, one or more aircraft in this category might have been expected to remain in production indefinitely, but this has not so far eventuated. The British Aerospace **Bulldog 100** was manufactured for several years during the 1970s, and a prototype of the retractable-gear Bulldog 200 or Bullfinch was built and flown in 1976, but this variant never reached production.

In all, 304 Bulldogs were constructed for eleven operators in nine nations. The first major export order was for the Swedish Air Force, although Sweden had its own primary trainer under development, in the form of the Saab **MFI-17 Supporter.** This sold to several countries, but in the late 1970s manufacture was transferred to Pakistan, where it was to be built under licence as the Mushshak by the Kamra Aeronautical Centre.

The de Havilland Tiger Moth. (Roy Braybrook)

Stampe SV.4. (Roy Braybrook)

Yakovlev Yak-11. (Roy Braybrook)

North American T-6 Harvard. (Roy Braybrook)

British Army de Havilland Chipmunk. (Roy Braybrook)

North American T-28
Trojan/Fennec. (Roy Braybrook)

Prototype of the Bulldog 200.
(Roy Braybrook)

Approximately 150 have been assembled or manufactured for the Pakistan services, and production reportedly continues at 15 per year. The introduction of a 210 hp turbocharged engine is under consideration. Total production of the Supporter amounts to well over 200 aircraft.

The third of the 200 hp primary trainers selling in the 1970s was the **CT-4 Airtrainer,** which was derived from Dr Millicer's Airtourer/ Aircruiser series by New Zealand Aerospace Industries. The project is now owned by Pacific Aerospace Corporation (PAC). The CT-4 is operated by the air forces of Australia, New Zealand and Thailand, and is being marketed by PAC in both piston- and turbine-engined forms, with fixed or retractable undercarriages.

Most piston-engined trainers in production fall in the 160-260 hp bracket. An engine of 160 hp now represents a realistic minimum for grading, hence the basic Trago Mills SAH-1 with 110 hp is underpowered for this role. The Slingsby T.67M Firefly is offered with either a 160 or 200 hp engine, while the FFA AS.202 Bravo is available with either a 150 or 180 hp unit. These aircraft, together with the 180 hp UTVA-75, are probably the least powerful types that should be considered for the primary training role.

The fixed-gear primary trainer with side-by-side seating continues up to the 235 hp Rallye 235G Guerrier, but around 260 hp there is a transition to retractable undercarriages, the only fixed-gear trainers in this class being the HPT-32 and AS.202/26A. The highly successful SF.260 is somewhat unusual in combining a retractable gear with side-by-side seating. Most aircraft at the upper end of the current piston-engined trainer range feature tandem seating. This is exemplified by the 290 hp IAR-831, 300 hp Epsilon, Pillán, Firecracker and Lasta, and the 325 hp PZL 130 Orlik.

For a long time during the 1970s it appeared that the piston-engined military trainer might disappear completely, since AVGAS was difficult to obtain in many parts of the world, and air forces wanted to standardise on turbine fuel. In recent years, however, the high cost of turbine engines is better appreciated, and the oil companies have taken steps to improve the distribution of AVGAS. In this new climate, piston-engined primary trainers are staging something of a come-back, although the introduction of modern technology reciprocating engines is being delayed by the recession in the general aviation industry.

Trago Mills SAH-1

The SAH-1 is a fully aerobatic two-seat primary trainer with a low wing and side-by-side seating. Initially developed with a 118 hp Avco Lycoming 0-235 engine, it is offered with a 160 hp unit for military purposes. The prototype (G-SAHI) first flew on 23 August 1983.

The SAH-1. (Trago Mills)

Powerplant: *160 hp Avco Lycoming AEIO-320.*
Dimensions: *wingspan 30 ft 8.4 in (9.36 m), length 21 ft 7^1/4 in (6.58 m), height 7 ft 9.6 in (2.38 m), wing area 120 sq ft (11.15 m^2).*
Weights: *empty equipped 1013 lb (450 kg), max fuel 188 lb (85 kg), max take-off 1648 lb (748 kg).*
Performance: *max level speed at sea level 140 knots (260 km/hr), sea level cruise speed at 75 per cent power 126 knots (233 km/hr), max rate of climb 1300 ft/min (6.6 m/sec), service ceiling 21,800 ft (6650 m), max range 500 nm (925 km) with 4 Imp gal (18 litres) reserves.*

Slingsby Aviation T67M Firefly

The Firefly is a two-seat primary trainer with side-by-side seating and a low wing. Its shape is based on that of the Fournier RF.6B, the original T67A (which first flew on 15 May 1981) being a licence-built RF.6B-120, retaining the wooden construction of the French original. Slingsby then redesigned the airframe in GRP (glass-reinforced plastic), and built the 113/160 hp civil T67B/C and the 160/200 hp military T67M Firefly. The T67M-160 first flew on 5 December 1982 and the T67M-200 on 16 May 1985. The T67 was the world's first fully aerobatic GRP light aircraft to achieve certification. Military testing has included a fatigue programme based on the Bulldog/Jet Provost spectrum, covering the equivalent of 75,000 flight hours, and giving a safe life of 15,000 hours. The Firefly has no external stores provisions. The description below relates to the T67M-200.

Powerplant: 200 hp Avco Lycoming AEIO-360.
Dimensions: wingspan 34 ft 9 in (10.6 m), length 23 ft 0 in (7.01 m), height 7 ft 9 in (2.36 m), wing area 136 sq ft (12.63 m²).
Weights: empty 1510 lb (685 kg), max fuel 252 lb (114 kg), max aerobatic/utility 2150 lb (975 kg).
Performance: max level speed at sea level 140 knots (259 km/hr), sea level cruise speed at 75 per cent power 136 knots (252 km/hr), max rate of climb 1150 ft/min (5.84 m/sec), range 480 nm (890 km) with 45 minutes reserves.

The T67M Firefly. (Slingsby Aviation)

FFA AS.202 Bravo

The AS.202 was a joint development by FFA (Flug-und Fahrzeugwerke AG) in Switzerland and SIAI-Marchetti in Italy, the first prototype making its maiden flight in early 1969. The Italian company later withdrew from the programme, and (after extensive redesign of the tail) FFA certificated the aircraft and produced it with both 150 and 180 hp engines. A prototype flew in 1978 with a 260 hp engine, but no production ensued. Most sales have been of the 180 hp AS.202/18A as described below.

Indonesian Air Force AS.202/18A Bravo. (FFA)

Powerplant: *180 hp Avco Lycoming AEIO-360.*
Dimensions: *wingspan 32 ft 0 in (9.75 m), length 24 ft 7 in (7.50 m), height 9 ft 3 in (2.81 m), wing area 149.2 sq ft (13.86 m²).*
Weights: *empty equipped 1543 lb (700 kg), max fuel 236 lb (107 kg), max take-off aerobatic 2095 lb (950 kg), max take-off utility 2315 lb (1050 kg).*
Performance: *max level speed at sea level 130 knots (241 km/hr), cruising speed at 75 per cent power at 8000 ft (2440 m) 122 knots (226 km/hr), max rate of climb 900 ft/min (4.6 m/sec), service ceiling 18,000 ft (5500 m), max range (no reserve) 520 nm (965 km).*

Valmet L-70 Vinka (Blast)

The L-70 was developed to provide the Finnish Air Force with a primary trainer. Design work began in 1970 and the first prototype had its maiden flight on 1 July 1975. A total of 32 were built, of which 30 were delivered to the service between 1980 and 1982, the other two being retained by Valmet for development work. The L-70 is offered for overseas sales under the name Miltrainer. It is more flexible in operation than most trainers in the 200 hp category, since the cockpit can accommodate up to four, and since it can be equipped with four pylons, two rated at 330 lb (150 kg) and two at 220 lb (100 kg). Valmet states that the L-70 has a fatigue life of 8000 hours in heavy military use.

Powerplant: 200 hp Avco Lycoming AEIO-360.
Dimensions: wingspan 31 ft 7 in (9.63 m), length 24 ft 7 in (7.50 m), height 10 ft 10 in (3.31 m), wing area 156 sq ft (14.5 m²).
Weights: empty equipped 1690 lb (767 kg), max internal fuel 269 lb (122 kg), max external load 660 lb (300 kg), max take-off aerobatic 2293 lb (1040 kg), max take-off normal 2755 lb (1250 kg).
Performance: max level speed at sea level 127 knots (235 km/hr), cruising speed at 75 per cent power at sea level 113 knots (210 km/hr), max rate of climb 1120 ft/min (5.7 m/sec), service ceiling 15,750 ft (4800 m), max range (no reserves) 525 nm (970 km).

Finnish Air Force L-70s. (Valmet)

Saab-Scania Supporter

As discussed previously, Saab built a number of MFI-17s under the name Supporter for military use, and as the Safari for civil use. Overseas customers for the Supporter have included Denmark and (recently) Norway. The Supporter is now produced in Pakistan under the Urdu name Mushshak (Proficient). It is understood that Pakistan has export rights, but at time of writing no overseas sales have taken place. The aircraft has six underwing hardpoints. The space at the rear of the cabin can accommodate a third man or 220 lb (100 kg) load. The original MFI-17 had its maiden flight on 26 February 1971.

Supporters of the Royal Danish Air Force. (A. Andersson, Saab-Scania)

Powerplant: 200 hp Avco Lycoming IO-360.
Dimensions: wingspan 29 ft 0 in (8.85 m), length 22 ft 11^{1}/$_{2}$ in (7.0 m), height 8 ft 6^{1}/$_{2}$ in (2.60 m).
Weights: empty equipped 1410 lb (640 kg), max external load 660 lb (300 kg), max take-off aerobatic 1985 lb (900 kg), max take-off weight normal 2645 lb (1200 kg).
Performance: max level speed at sea level 128 knots (235 km/hr), cruising speed at 75 per cent power at sea level 112 knots (208 km/hr), max rate of climb 810 ft/min (4.1 m/sec), service ceiling 13,500 ft (4100 m), max endurance with 10 per cent reserves 5 hr 10 min.

Powerplant: 200 hp Avco Lycoming IO-360.
Dimensions: wingspan 33 ft 0 in (10.06 m), length 23 ft 3 in (7.08 m), height 7 ft 6 in (2.28 m), wing area 129.4 sq ft (12.02 m²).
Weights: empty (with standard equipment) 1430 lb (645 kg), internal fuel 230 lb (104.5 kg), max take-off aerobatic 2238 lb (1015 kg), max take-off semi-aerobatic 2350 lb (1066 kg).
Performance: max level speed at sea level 130 knots (241 km/hr), max cruising speed at 75 per cent power at 4000 ft (1220 m) 120 knots (222 km/hr), max rate of climb 1034 ft/min (5.25 m/sec), service ceiling 16,000 ft (4880 m), max range (no allowances) 540 nm (1000 km).

The Bulldog 100 series, as flown by the University Air Squadrons. (British Aerospace)

British Aerospace Bulldog 100-Series

The Bulldog was originally designed by Beagle Aircraft, then taken over by Scottish Aviation, which later became part of British Aerospace. First flight dates were 19 May 1969 with Beagle and 14 February 1971 with Scottish Aviation. A crucial factor in the success of the Bulldog was its selection by the Swedish Air Force for primary flying training and army liaison duties. This requirement involved firing trials with tip-mounted Bofors wire-guided missiles, to clear the aircraft for use in GW training. The Bulldog has an exceptionally large cockpit to accommodate pilots in full flying kit, and a third seat. It has also been cleared for operation from skis. In RAF use it is operated primarily by the University Air Squadrons.

A prototype of the retractable-gear Bulldog 200 or Bullfinch was flown in 1976. Its empty weight was increased to 1810 lb (820 kg), its aerobatic weight to 2304 lb (1045 kg) and its maximum weight to 2601 lb (1180 kg). Maximum level speed at sea level was 150 knots (278 km/hr).

Pacific Aerospace Corporation CT-4 Airtrainer

Although not currently in production, the CT-4 is still being marketed, and is of special interest since it was described by its original manufacturer (New Zealand Aerospace Industries) as 'the most versatile and competitive basic trainer on the market today'. The CT-4 is certainly unusual in being designed to be fully aerobatic (6G) at maximum all-up weight. It also has a somewhat higher wing loading than is normal in this category, which improves the ride quality in turbulence, the effect on stall speed being offset by large, efficient flaps. Provisions are made for a third crew member and for tiptanks. The Airtrainer serves as the CT-4A in Australia, the CT-4B in Thailand, and the CT-4 in New Zealand. The CT-4 first flew on 21 February 1972. ▷

Powerplant: 210 hp Continental IO-360.
Dimensions: wingspan 26 ft 0 in (7.92 m), length 23 ft 2 in (7.06 m), height 8 ft 6 in (2.59 m), wing area 129 sq ft (12.0 m²).
Weights: empty equipped 1520 lb (690 kg), max take-off 2350 lb (1066 kg).
Performance: max level speed at sea level 154.5 knots (286.5 km/hr), cruising speed at 75 per cent power at sea level 140 knots (259 km/hr), max rate of climb 1350 ft/min (6.86 m/sec), service ceiling 17,900 ft (5460 mm), range with 10 per cent reserves 650 nm (1205 km) or 1200 nm (2225 km) with tiptanks.

New Zealand's CT-4, exhibited at Paris in 1975. (Roy Braybrook)

Pacific Aerospace has recently proposed the CT-4/CR, a new-build aircraft with an Allison 250-B17 turboprop engine and a retractable undercarriage. Dimensions are the same as for the CT-4, except that length is increased to 23.45 ft (7.15 m). Maximum take-off weight is 2450 lb (1111 kg). Performance for the CT-4/CR is estimated as follows: maximum speed 240 knots (445 km/hr), cruising speed at 75 per cent power 210 knots (389 km/hr), max rate of climb 3050 ft/min (15.5 m/sec), service ceiling 32,500 ft (9910 m), range at 75 per cent power 724 nm (1340 km). The CT-4/C, with the Allison engine but a fixed undercarriage, is also available.

FUS/MBB Flamingo

The MBB 223 Flamingo T-1 has been built in Germany, Switzerland and Spain under the control of Flugzeug-Union-Sued, an MBB subsidiary. The first flight took place on 2 March 1987. In all, just less than 100 were built, equipped with a 200 hp Avco Lycoming IO-360 engine, although FUS subsequently converted a demonstrator to take the 210 hp turbocharged TO-360.

At the Hanover Air Show in 1986 this demonstrator was presented with the TO-360 replaced by a Porsche PFM 3200. This was offered as a

Powerplant: 212 hp Porsche PFM 3200.
Dimensions: wingspan 27 ft 2¹/₂ in (8.28 m), length 25 ft 2¹/₂ in (7.68 m), height 8 ft 10¹/₂in (2.70 m), wing area 124 sq ft (11.5 m²).
Weights (for TO-360 version): empty equipped 1545 lb (700 kg), max fuel 270 lb (122.4 kg), max take-off weight 2315 lb (1050 kg).
Performance: maximum speed 133 knots (246 km/hr), cruising speed 119 knots (220 km/hr), max rate of climb 1000 ft/min (5.08 m/sec), service ceiling 15,000 ft (4560 m), max range 534 nm (990 km).

The FUS Flamingo trainer with the standard Avco Lycoming engine. (Flugzeug-Union-Sued)

conversion kit, although in the future there may be new-build Flamingos with the Porsche engine. This new engine has the advantage of being certificated for MOGAS, and of providing a 10-15 per cent improvement in fuel consumption. It also features single-lever operation, a fan-air cooling system, a geared propeller and a silencer, giving a 5 or 6 dB noise reduction, and a digital fuel consumption display.

Zlin Z-142

The Zlin 142. (Omnipol)

The Z-142 is a derivative of the Zlin 42M, the new aircraft making its first flight on 29 December 1978. Production began in 1981, and the aircraft has been exported to Cuba, East and West Germany, Hungary, Poland and Romania. The Z-142 is believed to be used by both military and civil operators in the primary-basic training role.

Powerplant: 210 hp Avia M 337 AK.
Dimensions: wingspan 30 ft 0^{1}/2 in (9.16 m), length 24 ft 0^{1}/2 in (7.33 m), height 9 ft 0 in (2.75 m), wing area 141.5 sq ft (13.15 m²).
Weights: empty 1610 lb (730 kg), internal fuel 198 lb (90 kg), max take-off aerobatic 2139 lb (970 kg), max take-off standard 2403 lb (1090 kg).
Performance: max level speed 125 knots (231 km/hr), max cruising speed 106 knots (197 km/hr), max rate of climb 1080 ft/min (5.5 m/sec), service ceiling 16,400 ft (5000 m), range at max cruising speed 283 nm (525 km), or 513 nm (950 km) with two 11 Imp gal (50 litre) auxiliary tanks.

SIAI-Marchetti SF.260

The origins of the highly successful retractable-gear SF.260 series may be traced back to the F.250, which was designed by Selio Frati and built by Aviamilano, making its first flight on 15 July 1964. The two/three seat military SF.260M flew on 10 October 1970, and this was followed by the SF.260W Warrior, which has provisions for up to four pylons. Well over 500 military SF.260s have been sold to around 15 operators, the largest sale being to Libya, which received a remarkable 240 aircraft. The turboprop SF.260TP is described separately. The SF.260W can carry two 275 lb (125 kg) bombs in combination with a single crew member and 330 lb (150 kg) of fuel.

Powerplant: *260 hp Avco Lycoming TO-540.*
Dimensions: *wingspan 27 ft 4¹/₂ in (8.35 m), length 23 ft 3¹/₂ in (7.10 m), height 7 ft 11 in (2.41 m), wing area 108.7 sq ft (10.1 m²).*
Weights: *empty equipped 1797 lb (815 kg), internal fuel including tiptanks 372.5 lb (169 kg), clean take-off 2513 lb (1140 kg), max take-off (SF.260W) 2866 lb (1300 kg).*
Performance: *max level speed at sea level 180 knots (333 km/hr), cruising speed at 75 per cent power at 4900 ft (1500 m) 162 knots (300 km/hr), max rate of climb 1500 ft/min (7.62 m/sec), service ceiling 15,300 ft (4665 m), max range 890 nm (1650 km).*

An SF.260MB of the Belgian Air Force. (SIAI-Marchetti)

Powerplant: *290 hp Avco Lycoming IO-540.*
Dimensions: *wingspan 32 ft 9½ in (10.0 m), length 29 ft 2½ in (8.9 m), height 7 ft 9½ in (2.38 m), wing area 161.4 sq ft (15 m²).*
Weights: *empty 2094 lb (950 kg), max take-off aerobatic 2646 lb (1200 kg), max take-off normal 3308 lb (1500 kg).*
Performance: *max level speed 172.7 knots (320 km/hr), cruising speed at 75 per cent power 159 knots (295 km/hr), max rate of climb 1380 ft/min (7.0 m/sec), ceiling 18,370 ft (5600 m), max range 700 nm (1300 km).*

IAR-831 Pelican

At the Paris Air Show of 1983 Romania showed for the first time this piston-engined derivative of the IAR-825TP turboprop trainer, which had flown only a few weeks prior to the show. It was stated that the -831 had been developed primarily for export, but that it might also be used domestically for sport flying and possibly by the Romanian Air Force. It was said to contain some elements of the IAR-823 two/four seater.

The IAR-831 Pelican in front of the old control tower at Le Bourget in 1985. (Roy Braybrook)

Norman Aeroplane Co. NDNIA Firecracker

The Firecracker began life with a 260 hp Avco Lycoming AEIO-540, and was effectively a tandem-seat rival to the well-established SF.260. The British aircraft has a very spacious cockpit, large ground clearance for its four underwing hardpoints, a ventral airbrake, and a low aspect ratio wing

that is claimed to demonstrate some of the handling characteristics of current operational aircraft. More recently the Firecracker has been offered with provisions for the Stencel Ranger assisted-escape system, and a 300 hp AEIO-540 (as described below). The Firecracker first flew on 26 May 1977. A turboprop derivative is described separately.

The prototype Firecracker (G-NDNI) with 260 hp engine (Roy Braybrook)

Powerplant: *300 hp Avco Lycoming AEIO-540.*
Dimensions: *wingspan 26 ft 0 in (7.925 m), length 25 ft 3 in (7.7 m), height 9 ft 11 in (3.02 m), wing area 128 sq ft (11.89 m²).*
Weights: *empty equipped 2163 lb (981 kg), internal fuel 598 lb (272 kg), max clean 3300 lb (1498 kg), max with external loads 3750 lb (1700 kg).*
Performance: *max level speed at sea level 167 knots (309 km/hr), cruising speed at 65 per cent power at 8000 ft (2440 m) 150 knots 278 (km/hr), max rate of climb 960 ft/min (4.88 m/sec), range 1100 nm (2038 km).*

Aérospatiale TB-30B Epsilon

The TB-30B Epsilon is a primary/basic trainer developed to act as a lead-in to the Alpha Jet in the French Air Force pilot training system. It first flew on 22 December 1979. After extensive redesign of the rear end, 150 were ordered for the French Air Force, with deliveries over the period 1983-88.

It is offered for export with four hardpoints, giving a total external load of up to 660 lb (300 kg). The first prototype is flying with the Turboméca TP319 engine derated to 350 shp. This Turbo-Epsilon was originally proposed for the RAF, and it remains to be seen if it will now be marketed.

The second prototype Epsilon over the Alps. (Aerospatiale)

Powerplant: *300 hp Avco Lycoming AEIO-540.*
Dimensions: *wingspan 25 ft 11$^{1}/_{2}$ in (7.92 m), length 24 ft 10$^{1}/_{2}$ in (7.59 m), height 8 ft 7$^{1}/_{2}$ in (2.63 m), wing area 96.9 sq ft (9.0 m^2).*
Weights: *empty equipped 2055 lb (932 kg), fuel 330 lb (150 kg), take-off aerobatic (6.7G) 2755 lb (1250 kg), with external stores 3087 lb (1400 kg).*
Performance: *max level speed at sea level 205 knots (380 km/hr), cruising speed at 75 per cent power at 6000 ft (1830 m) 193 knots (358 km/hr), max rate of climb 1850 ft/min (9.4 m/sec), service ceiling 23,000 ft (7010 m), max range 680 nm (1260 km).*

ENAER T-35 Pillán

In the late 1970s the Chilean Air Force planned to buy New Zealand's CT-4, but supply of this aircraft was banned for political reasons. Chile therefore established its own aircraft manufacturing facility, with the longer-term aim of developing types to fulfil local needs. IndAer assembled Piper PA-28 Dakotas for military and civil use, and in 1980 work began with Piper to develop a primary/basic trainer, using some existing components from the Dakota and PA-32 Saratoga. On 6 March 1981 the prototype of the new aircraft was flown by Piper, and production began in 1984 in Chile. At this stage IndAer was reformed as ENAER (Empresa Nacional de AERonautica). Deliveries of 80 of these aircraft, designated T-35 Pillán (Devil) began

in 1985, and Spain later ordered 40 T-35Bs for assembly by CASA. The Spanish Air Force designation is E-26 Tamiz ('Sieve').

Powerplant: 300 hp Avco Lycoming AEIO-540.
Dimensions: wingspan 28 ft 11 in (8.81 m), length 26 ft 1½ in (7.97 m), height 7 ft 8½ in (2.34 m), wing area 147 sq ft (13.62 m²).
Weights: empty 1836 lb (833 kg), fuel 432 lb (196 kg), max take-off clean 2900 lb (1315 kg).
Performance: max level speed at sea level 168 knots (311 km/hr), cruising speed at 75 per cent power at 8800 ft (2680 m) 161 knots (298 km/hr), max rate of climb 1525 ft/min (7.75 m/sec), service ceiling 19,100 ft (5820 m), range with 45 min reserves 625 nm (1157 km).

Chilean Air Force T-35 Pillan.
(ENAER)

The UTVA Lasta. (Swallow)
(SDPR)

UTVA Lasta

Yugoslavia's UTVA company exhibited a model of its new NKA (new generation trainer) at the Paris Air Show of 1985, a few weeks before the prototype first flew. In production form the aircraft is known as the Lasta (Swallow), and is very similar to France's Epsilon. The cockpit of the Lasta is designed to be as close as possible to that of the Soko G-4 Super Galeb, to which Yugoslav Air Force students will graduate for advanced flying training. The Lasta has two hardpoints, designed for a total load of up to 880 lb (400 kg).

Powerplant: 300 hp Avco Lycoming AEIO-540.
Dimensions: wingspan 27 ft 4$\frac{1}{2}$ in (8.34 m), length 26 ft 4$\frac{1}{2}$ in (8.04 m), height approx 10 ft 2$\frac{1}{2}$ in (3.1 m), wing area 118.4 sq ft (11.0 m^2).
Weights: empty equipped 2337 lb (1060 kg), max take-off 3594 lb (1630 kg).
Performance: max level speed at sea level 186 knots (345 km/hr), max rate of climb 1770 ft/min (9.0 m/sec).

PZL-130 Orlik

Poland's Orlik (Eaglet) is unique in the sense that it is the only new trainer design in recent years to feature a radial engine. It is also unusual in applying a system approach normally found only in high performance jet trainers. In parallel with the aircraft itself, this branch of PZL has also developed the PZL-130 Professor flight simulator and the PZL-130 Inspector diagnostic equipment for the aircraft systems. The Orlik first flew on 12 October 1984, and a prototype appeared at the Paris Air Show in the following year. The cockpit of this aircraft is generally similar to that of the TS-11 Iskra (Spark) jet trainer, to facilitate conversion.

The PZL-130 Orlik. (Pezetel)

Powerplant: 325 hp Vedeneyev M-14Pm.
Dimensions: wingspan 26 ft 3 in (8.0 m), length 27 ft 8½ in (8.45 m), height 13 ft 1½ in (4.0 m), wing area 132.4 sq ft (12.3 m²).
Weights: empty equipped 2447 lb (1110 kg), fuel 683 lb (310 kg), max take-off aerobatic 3196 lb (1450 kg), max take-off 3527 lb (1600 kg).
Performance: max level speed 197 knots (365 km/hr), max cruising speed 174 knots (322 km/hr), max rate of climb 1575 ft/min (8.0 m/sec), service ceiling 17,000 ft (5200 m), max range 765 nm (1420 km).

Yakovlev Yak-52

The Yak-52 is a replacement for the Yak-18, which first flew in 1946. Existence of the new aircraft was announced in 1978, and in the following year production was allocated to Romania, where more than 500 are reported to have been built. An interesting feature of the Yak-52 is its semi-retracting main-wheels, which project below the wing in order to minimize damage to the airframe in the event of a forced landing.

Powerplant: 360 hp Vedeneyev M-14P radial engine.
Dimensions: wingspan 30 ft 6½ in (9.3 m), length 25 ft 5 in (7.745 m), height 8 ft 10½ in (2.7 m), wing area 161.5 sq ft (15.0 m²).
Weights: empty 2205 lb (1000 kg), fuel 220 lb (100 kg), max take-off 2844 lb (1290 kg).
Performance: max level speed 162 knots (300 km/hr), max cruising speed 145 knots (270 km/hr), max rate of climb 1380 ft/min (7.0 m/sec), service ceiling 19,700 ft (6000 m), max range 300 nm (550 km).

The Romanian-built Yak-52. (Roy Braybrook)

Pilatus PC-9 demonstrator at Farnborough 1986. (Roy Braybrook)

A PC-9 for the Royal Saudi Air Force. (Pilatus Aircraft)

Two prototypes and two production examples of the Fairchild Republic T-46A flew before the programme was halted. (Fairchild)

Above: A rare shot of the Aero L-39 Albatross in desert camouflage and Czech markings. (Omnipol)

Below: The Israel Aircraft Industries AMIT (Advanced Multimission Improved Trainer) *Tzukit* (Thrush) is a refurbished and modernised Magister. (IAI)

Below: C-101 armed with what appear to be four napalm tanks and a twin 12.7 mm machine gun pod. (CASA)

Centre: Illustrating the anti-shipping strike potential of the Alpha Jet, this NGEA is armed with an AM.39 Exocet missile and two Matra Magic air-air missiles. (Dassault-Breguet)

Above: Nine MB-339PANs of the *Pattuglia Acrobatica Nazionale,* the Italian Air Force aerobatic team, inverted in the course of their display. (Aermacchi)

Below: An Egyptian Air Force Alpha Jet in landing attitude. The pointed nose houses a laser ranger, indicating this is the MS2 variant, which is now being brought to NGEA standard. (Dassault-Breguet)

Above: The MB-339C, equipped with full nav-attack system and armed with 30 mm DEFA cannon pods, rocket pods and Sidewinders. Note the enlarged cylindrical tiptanks. (Aermacchi)

Above: The CASA C-101DD with nose radome and ventral 30 mm cannon pod, at Farnborough 1986. (Roy Braybrook)

Centre: A British Aerospace Hawk Mk53, photographed on a delivery flight to Indonesia. (Trevor Davies, BAe)

Below: The Soko G-4 Super Galeb, making its public debut in Paris in 1983. (Roy Braybrook)

The Hawks of the RAF Red Arrows team, pictured during a demonstration tour of the Middle East. (Geoffrey Lee, BAe)

Inset: This full-scale plastic replica of the McDonnell Douglas T-45 was produced by Specialised Mouldings of Huntingdon, and was exhibited by BAe at Farnborough in 1984. (Roy Braybrook)

Chapter 6

Turboprop Trainers

THE TURBOPROP basic trainers that emerged in the 1970s benefited from the availability of engines developed for commercial applications, in the same way that the first purpose-built basic jet trainers owed their existence to the availability of what was then a new generation of lightweight turbojets, several of which had been developed for missile or drone applications (as discussed in Chapter 2).

The most important turboprop in this context has so far been the **Pratt & Whitney Canada (P&WC) PT6A,** an engine that first ran on the bench in November 1959, and flew in the nose of a Beech 18 in May 1961. At the start of 1986 over 19,500 PT6As had logged almost 100 million hours in 9157 aircraft registered in 144 countries. The PT6A equips a vast array of light transport aircraft, such as the Cessna Conquest and Caravan series, Beech King Air and 1900, the Piper Cheyenne, the IAI Arava, DHC Twin Otter and Dash-7, Fairchild Metro IIIA, Embraer Bandeirante and Xingu, Shorts 330 and 360, Pilatus PC-6, and various turboprop conversions of agricultural aircraft.

The **Allison 250** series of engines came in at a later stage, filling the gap between the PT6A and the top of the piston range. The Allison turboprop powers the Nomad and Turbo-Islander light transports, but it is best known as a turboshaft engine for helicopter applications, including the Agusta A.109, Aerospatiale AS.355, MBB NO 105, MD.500 series, the Sikorsky S-76, and a wide range of Bell products. At the start of 1986 more than 21,000 Allison 250s had been delivered. As mentioned earlier, in the 1990s this engine may find itself in competition with both the Turboméca TP319 and the Teledyne Continental TP500.

The Garrett TPE331 series entered the trainer market only in the mid-1980s, as a result of Shorts looking for an alternative to the PT6A-62 for the Tucano in competing with the PC-9. The 331 series was developed for the OV-10 Bronco (as the T76) and a range of civil applications, including the Mitsubishi MU-2 family, Turbo Commander, Shorts Skyvan, Fairchild Merlin and Metro, King Air B100, CASA C-212, Dornier Do228, Cessna Conquest II, BAe Jetstream, and Piper Cheyenne 400LS. At the time of the AST.412 contest, Pilatus argued that the TPE331 is unsuitable for training aircraft, since its single-shaft layout gives a fast response that is not representative of most turbine engines. On the other hand, the RAF appears to have taken the view that this response gives a useful safety margin, and that it could be slowed by a simple fuel system modification if this proved necessary. At the start of 1986 deliveries of the TPE331 had passed 10,000 units, and engine service hours stood at over 32 million.

The background to the three main engine types has been discussed in some detail in order to make the point that this new generation of relatively low cost basic trainers was made possible by civil demands for turboprops to power light twins and commuter aircraft. Fortuitously, these commercial engines required very little modification to suit 6G trainers, beyond the introduction of lubrication and fuel system changes to permit inverted flight. Aside from the comparatively low unit cost that results from a high production rate, these engines provided from the outset a comparatively long time between overhauls (TBO), with consequent benefit to operating costs.

Development costs and technical problems have certainly been minimised by the availability of well-proven, reliable engines, but several turboprop trainers have suffered difficulties in achieving satisfactory handling characteristics, especially in the context of spinning. The Beech T-34C, when subjected to preliminary evaluation by the US Navy, was criticised for its spin and spin recovery characteristics, which took a seven-month development programme to correct. The Valmet L-80TP was lost in a spinning accident, requiring considerable redesign of the tail surfaces. The prototype Turbo-Orlik was lost in a fatal accident after only five months of flight testing, although the cause has not yet been published.

In contrast, the Pilatus PC-7 which blazed the trail for turboprop trainers, appears to have enjoyed a comparatively trouble-free development. The problem for the PC-7 was that it was a decade ahead of the market demand.

The market for the turboprop trainer was slow to develop, and it has eventuated in two phases, the first relating mainly to the Third World, and the second to the major air forces. Although developed for the US Navy, the T-34C has sold to Angola, Argentina, Ecuador, Gabon, Indonesia, Morocco, Peru and Uruguay. Likewise, the PC-7 has been exported to Abu Dhabi, Angola, Austria, Bolivia, Burma, Chile, Guatemala, Iran, Iraq, Malaysia, and Mexico. The Swiss order was placed in 1981, three years after export deliveries had begun. It may be that the Swiss order represented the leading edge of the second wave of turboprop sales, with

Australia ordering the PC-9, and Britain the Shorts Tucano in the mid-1980s.

Turboprops nonetheless have a long way to go in terms of sales to major air forces. The USAF does not use turboprops, nor (currently) do the air forces of France, West Germany, Israel, Italy, Japan, and Sweden. In spite of this limited penetration of the top end of the market, during the 10-year period 1977-86 the turboprop trainer has expanded to provide a broad spectrum of products from 350 to 1100 shp. The following survey discusses the principal types roughly in ascending order of power. It thus deals firstly with the aircraft equipped with the Allison 250 series, then the P&WC PT6A, and finally the Garrett TPE331.

SIAI-Marchetti SF.260TP

SIAI-Marchetti SF-260TP. (Roy Braybrook)

This development of the SF.260M/W represents the simplest possible change from a piston to a turboprop engine, since the new aircraft is virtually identical to the original aft of the firewall. This makes conversion in the field very straightforward, but it results in limited range and

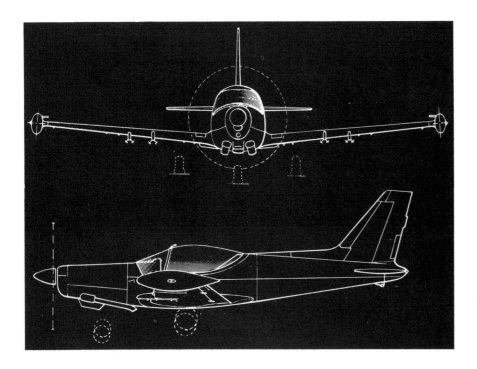

General arrangement drawing of the SF.260TP. (SIAI-Marchetti)

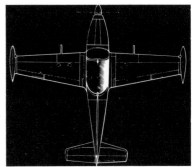

Powerplant: Allison 250-B17D flat-rated at 350 shp.
Dimensions: wingspan 27 ft 4¹/₂ in (8.35 m), length 24 ft 3¹/₂ in (7.40 m), height 7 ft 11 in (2.41 m), wing area 108.7 sq ft (10.1 m²).
Weights: empty equipped 1654 lb (750 kg), internal fuel including tiptanks 420 lb (190 kg), max take-off (trainer) 2645 lb (1200 kg), max take-off (armed) 2866 lb (1300 kg).
Performance: max level speed at sea level 206 knots (382 km/hr), at 10,000 ft (3000 m) 230 knots (426 km/hr), cruise speed at 10,000 ft (3000 m) 216 knots (400 km/hr), max rate of climb 2170 ft/min (11 m/sec), service ceiling 25,000 ft (7600 m), max range 512 nm (950 km) with 30 min reserves.

endurance, since a turbine engine is considerably more thirsty than the reciprocating variety.

The only modifications aside from the powerplant are the introduction of a trim tab on the rudder and an automatic fuel feed system. The aircraft is designed to achieve a safe life of 8000 flying hours and 20 years of operation.

The prototype SF.260TP flew on 1 July 1980 and deliveries began in 1982. At time of writing reports indicate that more than 50 new-build aircraft have been sold and that various countries (eg, Zimbabwe) are converting SF.260s in the field.

Powerplant: Allison 250-B17D flat-rated at 360 shp.
Dimensions: wingspan 33 ft 11 in (10.34 m), length 25 ft 11 in (7.90 m), height 9 ft 4 in (2.85 m), wing area 158.75 sq ft (14.75 m²).
Weights: empty equipped 1962 lb (890 kg), internal fuel 650 lb (296 kg), max external load 1764 lb (800 kg), max take-off aerobatic 2980 lb (1350 kg), utility 3240 lb (1470 kg), normal 3530 lb (1600 kg), utility with external loads 4190 lb (1900 kg).
Performance: max level speed at 5000 ft (1500 m) 181 knots (335 km/hr), cruising speed at 75 per cent power at 10,000 ft (3000 m) 165 knots (305 km/hr), max rate of climb 1930 ft/min (9.8 m/sec), engine-limited service ceiling 25,000 ft (7600 m), endurance 5 hr plus, max range 810 nm (1500 km).

Prototype of the L-90TP Redigo. (Valmet)

Valmet L-90TP Redigo

When the Finnish Air Force retired the Magister basic jet trainer, this left a large performance gap between the piston-engined L-70 and the BAe Hawk. This situation encouraged Valmet to develop a turboprop derivative of the L-70, which was given a new wing, a retractable undercarriage, and a revised tail. The resulting L-80TP had its maiden flight on 12 February 1985. It was intended to show the aircraft at Paris in June, but it crashed in a spinning accident on 24 April on its fourteenth flight.

After redesign of the tail surfaces to correct the spinning characteristics, the aircraft was redesignated L-90TP Redigo, and in this form made its first flight on 1 July 1986. Like the L-70, the L-90TP has a large cockpit, able to seat up to four. It is unusual in being stressed for up to +7G and −3.5G, rather than +6G and −3G. The L-90TP is designed for a safe life of 10,000 hours and 30,000 landings. Six underwing hardpoints are provided. An alternative composite wing is being developed.

Hindustan Aeronautics HTT-34

The HTT-34 is a turboprop derivative of the piston-engined HPT-32, which has a 260 hp Avco Lycoming AEIO-540 engine, first flew on 6 January 1977, and is now in service with the Indian Air Force as a primary trainer. The first HTT-34 was in fact the third HPT-32, which flew with the Allison engine on 17 June 1984 and appeared at the Farnborough Air Show that September.

As in the case of the SF.260TP, the turboprop version differs from the piston-engined original only forward of the firewall. The HTT-34 is unique among turboprop trainers in having a fixed undercarriage, but (like the HPT-32) it is described as being capable of modification to a retractable gear. It remains to be seen whether the Indian Air Force will adopt this aircraft, with either undercarriage arrangement.

Powerplant: *Allison 250-B17D, flat-rated at 420 shp.*
Dimensions: *wingspan 31 ft 2 in (9.50 m), length 26 ft 5$\frac{1}{2}$ in (8.07 m), height 9 ft 5$\frac{1}{2}$ in (2.88 m), wing area 161.4 sq ft (15.00 m²).*
Weights: *empty 1910 lb (866 kg), internal fuel 382 lb (173 kg), max take-off weight 2690 lb (1220 kg).*
Performance: *max level speed at sea level 167 knots (310 km/hr), max rate of climb 2132 ft/min (10.83 m/sec), service ceiling 25,000 ft (7600 m), max range 332 nm (615 km), endurance 3 hr.*

Hindustan Aeronautics HTT-34 at Paris in 1985. (Roy Braybrook)

Powerplant: Allison 250-B17D, flat-rated at 420 shp.
Dimensions: wingspan 28 ft 11 in (8.81 m), length 27 ft 2¹/₂ in (8.29 m), height 7 ft 8¹/₂ in (2.34 m), wing area 147 sq ft (13.62 m²).
Weights: basic empty 2310 lb (1048 kg), internal fuel 507 lb (230 kg), max take-off 3000 lb (1364 kg).
Performance: max level speed at sea level 198 knots (367 km/hr), max cruising speed at 10,000 ft (3000 m) 186 knots (345 km/hr), max rate of climb 1930 ft/min (9.8 m/sec), engine-limited service ceiling 25,000 ft (7600 m), range 620 nm (1150 km).

Prototype of the Aucan in flight.
(ENAER)

ENAER Aucan

After completing development of the piston-engined Pillán, Chile's ENAER began the design of a turboprop derivative, which was initially designated Turbo-Pillán. The use of an Allison 250 was a natural choice, not only because it provided a suitable power increase over the 300 hp of the Pillán, but also in view of the fact that ENAER planned to undertake licence-production of the MBB Bo 105 helicopter, which is equipped with this engine.

The new project was later designated Aucan (Little Demon or Gremlin). It flew on 14 February 1986 and appeared at the FIDA-86 show in Chile during the following month. Production is planned for 1988.

Rhein-Flugzeugbau Fantrainer 400

The RFB company is now part of MBB, Germany's largest aerospace group. It was founded in 1956, and has been one of the leaders in the use of composite materials, the ram wing (surface effect vehicle), and the use of ducted fans. This type of propulsion was used for the Sirius powered glider, which first flew on 5 July 1971, then on the two-seat Fanliner light aircraft, which flew on 8 October 1973. Although the Fanliner did not enter series production, it demonstrated the potential of the ducted fan, and in March 1975 RFB was awarded a Defence Ministry contract to develop a tandem-seat basic trainer based on this concept.

The first Fantrainer prototype (AWI-2) was powered by two 150 hp Wankel engines, and had its maiden flight on 27 October 1977. The second had a 420 shp Allison 250-C20B turboshaft engine, and flew on 31 May 1978. The same aircraft in June 1980 flew with the 650 shp Allison 250-C30 engine. The aircraft with the -C20B and -C30 became the Fantrainer 400 and 600 respectively.

The front and centre fuselage sections consist of GRP (glass-reinforced plastic) fairings on a metal keel. Since the ducted fan is located in the centre fuselage, the tail loads have to be carried by four longerons running aft from the fan duct. Joining these longerons with shear webs produces a rear fuselage of cruciform cross-section.

In August 1982 Thailand signed a contract for 31 Fantrainer 400s and sixteen 600s, with an option on a further twenty-six 600s for the FAC role, replacing the 0-1 and OV-10. As designed by RFB, the Fantrainer has GRP wings, but the wings produced in Thailand (by the Royal Thai Air Force) are of conventional metal construction and are thus somewhat heavier. The first two aircraft were delivered to Thailand complete, but the remainder are being assembled locally, with the Thai content gradually increasing to 50 per cent.

It may be noted that Thailand has export rights for the other ASEAN countries, and that one Fantrainer 600 has been modified by the RTAF to take a General Electric M197 three-barrel 20 mm Gatling Gun with 500 rounds of ammunition.

In 1985 three Fantrainers (one 400 and two 600s) were evaluated by the German Air Force in a 240-hour flight test programme, which led to various changes, including a quieter fan. It is anticipated that Fantrainers may be purchased to replace the P.149D primary trainer in the late 1980s.

Press reports have recently indicated that RFB is considering the use of engines in the 1000 shp category, presumably to compete with aircraft such as the PC-9 and Shorts Tucano.

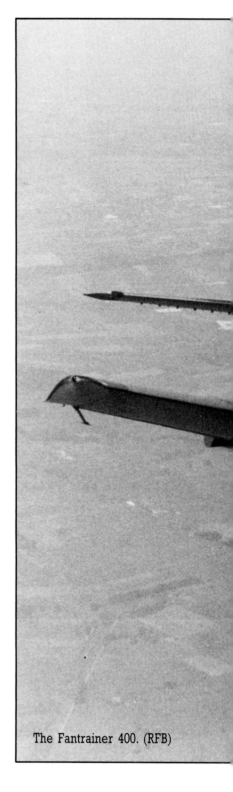

The Fantrainer 400. (RFB)

This rear-quarter view of the Fantrainer illustrates the cruciform rear fuselage. (RFB)

FT600 – Specification
This aircraft differs from the FT400 only as follows:
Powerplant: 650 shp Allison 250-C30.
Weights: empty 2557 lb (1160 kg), max take-off (utility) 5070 lb (2300 kg).
Performance: max level speed at 18,000 ft (5500 m) 225 knots (417 km/hr), cruising speed at 10,000 ft (3000 m) 200 knots (370 km/hr), max rate of climb 3000 ft/min (15.24 m/sec), service ceiling 25,000 ft (7600 m), max range 750 nm (1390 km), max endurance 4 hr 42 min.

FT400 – Specification
Powerplant: 420 shp Allison 25-C20B.
Dimensions: wingspan 31 ft 10 in (9.7 m), length 31 ft 1^1/$_2$ in (9.48 m), height 9 ft 10 in (3.00 m), wing area 150.7 sq ft (14.00 m²).
Weights: empty 2456 lb (1114 kg), max fuel (aerobatic) 418 lb (190 kg), max fuel 750 lb (340 kg), max take-off (aerobatic) 3527 lb (1600 kg), max take-off (utility) 3968 lb (1800 kg).
Performance: max level speed at 10,000 ft (3000 m) 200 knots (370 km/hr), cruising speed at 10,000 ft (3000 m) 175 knots (325 km/hr), max rate of climb 1550 ft/min (7.88 m/sec), service ceiling 20,000 ft (6100 m), max range 950 nm (1760 km), max endurance 6 hr 18 min.

Beech T-34C Turbine Mentor

The T-34C is a turboprop derivative of the US Navy's T-34B Mentor, which was powered by a 225 hp Continental 0-470 piston engine. The Mentor was a tandem-seat trainer based on the Bonanza four-seat cabin monoplane. Originally designated Model 45, the YT-34 first flew on 2 December 1948. The Mentor entered service in 1954 as the T-34A for the USAF and the T-34B for the US Navy. Beech built 450 T-34As and 423 T-34Bs, in addition to several hundred export models for Latin America and Spain. Some 126 were manufactured by Fuji in Japan, and 90 were assembled in Argentina.

In March 1973 the US Navy gave Beech a contract to design a turboprop version of the T-34B with updated avionics. Two YT-34C prototypes were built, the first having its maiden flight on 21 September 1973. The production order was delayed by problems in achieving acceptable spinning and spin recovery characteristics, but the initial contract was placed in mid-1975, with deliveries beginning in June 1976. In Navy service the T-34C replaced both the T-34B and the 800 hp North American T-28 Trojan. It may be added that in the meantime the USAF had abandoned propeller-driven trainers, replacing the T-34A with the Cessna T-37. By April 1984 the US Navy had received a total of 334 T-34Cs.

The speed with which the T-34C was produced was due in part to the use of existing Beechcraft components. The fuselage aft of the firewall remained that of the T-34B, while the wings were the outer panels of the Baron twin, the undercarriage and ailerons were taken from the Duke B60, and the engine was basically that of the King Air.

The P&WC PT6A-25 is theoretically capable of producing 680 shp, since the gas producer section is that of the PT6A-27, which has a larger compressor than preceding variants. However, the propeller reduction gearbox is basically that of the PT6A-20, which is restricted to 550 shp. All PT6A-25s are tested to 550 shp before leaving the factory, but those delivered to the US Navy are effectively derated to 400 shp on installation, by means of a special torque controller and fuel control unit. On a standard day this 400 shp (a rating that was chosen to increase engine life) is available up to an altitude of approximately 18,000 ft (5500 m).

Powerplant: P&WC PT6A-25, torque-limited to 400 shp.
Dimensions: wingspan 33 ft 7½ in (10.25 m), length 28 ft 10 in (8.79 m), height 9 ft 8½ in (2.96 m), wing area 179.6 sq ft (16.69 m²).
Weights: empty 2605 lb (1182 kg), empty equipped 2940 lb (1333 kg), internal fuel 840 lb (381 kg), max take-off 4300 lb (1950 kg).
Performance: max cruising speed at sea level 185 knots (343 km/hr), at 17,500 ft (5350 m) 226 knots (419 km/hr), max rate of climb 1430 ft/min (7.27 m/sec), service ceiling over 20,000 ft (6100 m), max range 795 nm (1475 km) with 5 per cent reserves and fuel for 20 min loiter.

One of two Beech YT-34C prototypes, serial 140861, as exhibited at Paris in 1975. (Roy Braybrook)

Beech T-34C-1 Turbine Mentor

The export version of the T-34C differs in exploiting the full 550 shp capability of the engine, and in having armament provisions. The front cockpit is equipped with a CA-513 reflector sight, and the aircraft has provisions for four underwing pylons, the inboard pair stressed for 600 lb (272 kg) loads on either side, and the outer pair for 300 lb (136 kg) loads. Maximum total external load is 1200 lb (544 kg). Whereas the standard aircraft has a usable fuel capacity of 107.5 Imp gal (490 litres), Beech offers a capacity of 125 Imp gal (568 litres) as an option. As in the case of the basic T-34C, the airframe is designed for a safe life of 16,000 flight hours and 30,000 landings.

Aside from the T-34C-1, Beech offers the Turbine Mentor 34C, which is understood to have the 550 shp rating but not the armament provisions. At time of writing export sales stand at 89 T-34C-1s and six Model 34Cs (for the Algerian MoD).

The T-34C-1 differs from the US Navy aircraft as follows:
Powerplant: *P&WC PT6A-25, flat-rated at 550 shp, which on a standard day is available to 7000 ft (2150 m).*
Weights: *empty equipped (trainer) 3000 lb (1360 kg), (armed) 3150 lb (1430 kg), max take-off (clean) 4300 lb (1950 kg) (armed) 5500 lb (2495 kg).*
Performance: *max cruising speed at 10,000 ft (3000 m) (clean) 230 knots (426 km/hr), (armed) 206 knots (382 km/hr), max rate of climb (clean) 2110 ft/min (10.72 m/sec), (armed) 1431 ft/min (7.27 m/sec), range (clean) 720 nm (1335 km), (with two 41.7 Imp gal (189 litre) (external tanks) 1150 nm (2130 km).*

A T-34C-1 Turbine Mentor of the
Royal Moroccan Air Force.
(Beech)

Pilatus PC-7 Turbo-Trainer

The PC-7 is a turboprop derivative of the 260 hp piston-engined P-3
trainer, which first flew on 3 September 1953. Having developed the 465 hp
P-2 in the mid-1940s and the P-3 in the mid-1950s, Pilatus decided to
develop a turboprop trainer in the mid-1960s. The prototype P-3 was
accordingly withdrawn from storage, redesignated P-3-06, and fitted with a
PT6A-20 with special modifications to permit inverted flight and aerobatics.
This was a bold move, since the original PT6A-6 had been certified only
in December 1963, and the -20 only in October 1965. Pilatus was laying the
foundations for a brand-new category of trainer, and it is probably true to

say that the aircraft that resulted was the best of the first turboprop generation.

The P-3 airframe was strengthened in critical areas to allow the performance potential of the new engine to be exploited, and provisions were made for the installation of tiptanks, increasing fuel volume from 35 to 90 Imp gal (159 to 409 litres). First flight took place on 12 April 1966, and the aircraft was exhibited at Paris in the following year.

Unfortunately for Pilatus, the general feeling at the time was that the trend was toward jet-powered trainers, which gave the smoothest transition to operational aircraft. It also became clear that there would be no short-term domestic market for what had now become the PC-7, since the Swiss Air Force intended to retain the P-3 into the 1980s. To make matters worse, the prototype was badly damaged in a wheels-up landing, following fuel system mismanagement by an evaluation pilot. Pilatus then shelved the PC-7 and reviewed its future projects.

Fortuitously, several changes soon occurred in the aerospace business, making turboprop trainers far more attractive. Rapid cost escalation had begun in the 1960s and continued into the 1970s, making air forces economise by finding less expensive ways to train pilots. In addition, the sudden increases in fuel prices in the early 1970s focussed attention on types of powerplant that emphasised propulsive efficiency.

A PC-7 Turbo-Trainer over the Alps. (Pilatus)

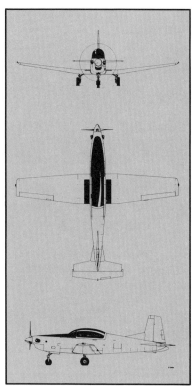

General arrangement drawing of the PC-7. (Pilatus)

Front cockpit of the PC-7. (Pilatus)

Powerplant: P&WC PT6A-25A, flat-rated at 550 shp.
Dimensions: wingspan 34 ft 1 in (10.40 m), length 32 ft 1 in (9.775 m), height 10 ft 6 in (3.21 m), wing area 179 sq ft (16.6 m²).
Weights: empty 2932 lb (1330 kg), internal fuel 847 lb (384 kg), max take-off (aerobatic) 4189 lb (1900 kg), (utility) 5952 lb (2700 kg).
Performance: max cruising speed at 20,000 ft (6100 m) 222 knots (412 km/hr), max rate of climb 2200 ft/min (11.2 m/sec), service ceiling 32,000 ft (9750 m), max range with 5 per cent reserves and fuel for 20 min loiter 647 nm (1200 km), or 1420 nm (2630 km) with auxiliary tanks, max endurance 4 hr 22 min.

Aside from the immediate problem of paying for fuel, the common perception at the time was that the world's known reserves of oil would all be expended within the life of the next generation of aircraft, hence fuel prices were apparently set to continue increasing.

Responding to this new interest, Pilatus leased a P-3 from the Swiss Air Force and modified it to take a PT6A engine. Thus powered it flew in May 1975. By this stage P&WC had developed the fully aerobatic PT6A-25, which Pilatus initially used derated to 515 shp, though it was later increased to 550 shp. The same rating is used in the PT6A-25A of the production PC-7, this modified engine having some magnesium castings to save weight. In the -25A provisions for pitch reversal and a water-wash spray-ring are deleted for the same reason.

This new prototype was presented at Paris in 1975, and was then subjected to a series of modifications to improve spinning behaviour. The wing was redesigned to provide a fuel capacity of 106 Imp gal (480 litres) in integral tanks, and six underwing hardpoints. Dihedral angle was increased, and the structure was restressed to permit an increase in take-off weight. The cockpit enclosures were redesigned to minimise the framework. The airframe was designed for a life of 12,000 hours and 24,000 landings.

Production appears to have been launched on the basis of only two firm orders: 12 for Burma and 8 for Bolivia. The first production aircraft flew on 18 August 1978, and was delivered early in the following year. At time of writing, orders stand in excess of 380 units, and the PC-7 continues to be offered for sale in parallel with the much more powerful PC-9. One recently introduced option (it was first offered at Paris in 1985) is the installation of lightweight Martin-Baker Mk15 ejection seats, giving escape on the runway at 60 knots (119 km/hr) and in flight up to 300 knots (556 km/hr).

Norman Aeroplane Co. NDNIT (Turbo) Firecracker

As discussed earlier, the Firecracker was developed by NDN Aircraft, which had been founded by 'Des' Norman in 1977. After its maiden flight on 26 May 1977, the company built a turboprop derivative, the NDNIT, which is sometimes referred to as the Turbo-Firecracker, and which first flew in September 1983.

Following the issuing of the first draft of the RAF's AST.412 for a Jet Provost replacement, in September 1983 a new company, Hunting Firecracker Aircraft, was formed specifically to promote and (if successful) to produce the NDNIT for the RAF. The aircraft had flown with a 550 shp PT6A-25A, but in order to meet the RAF demand for a speed of approximately 240 knots (445 km/hr) at sea level it was offered with a 750 shp PT6A-25D. This was similar to the -25C of the Embraer Tucano, but with aluminium castings in place of magnesium. The minimum speed capability that the RAF would accept was 210 knots (389 km/hr), and

Powerplant: P&WC PT6A-25A flat-rated at 550 shp.
Dimensions: wingspan 26 ft 0 in (7.925 m), length 27 ft 4 in (8.33 m), height 10 ft 8 in (3.25 m), wing area 128 sq ft (11.89 m²).
Weights: empty equipped 2552 lb (1160 kg), internal fuel 664 lb (301 kg), clean take-off weight 3600 lb (1636 kg), max take-off 4140 lb (1881 kg), max external load permissible with full fuel 757 lb (344 kg).
Performance: max level speed at sea level 196 knots (363 km/hr), at 5000 ft (1500 m) 203 knots (376 km/hr), economical cruising speed at 20,000 ft (6100 m) 180 knots (333 km/hr), max rate of climb 2060 ft/min (10.47 m/sec), service ceiling 27,100 ft (8260 m), max range on internal fuel 625 nm (1158 km), with two 32 Imp gal (145 litre) tanks 1100 nm (2040 km).

The NDNIT Turbo-Firecracker in flight. (Norman Aeroplane Co)

Hunting Firecracker predicted for the NDNIT a speed of 223 knots (413 km/hr). The NDNIT was promoted as the only British contender in the AST.412 contest, but in November 1984 it was dropped from the shortlist on cost and performance grounds.

Following the conclusion of the AST.412 affair, all rights, tools and jigs for the Firecracker have reverted to the Norman Aeroplane Company, which is offering the NDNIT with a 550 shp PT6A-25A. Although this aircraft may not have suited RAF needs, it is felt by some experts to provide in certain respects a realistic introduction to fast jet flying, its low aspect ratio wing making the student aware of angle of attack effects and the high sink rates that can be generated by swept wings.

For weapons training and light attack duties the NDNIT may be fitted with four pylons and a Singlepoint HUD Model SP.800. Customer options include the Stencel Ranger zero-zero pilot extraction system.

PZL-130T Turbo-Orlik

When the 300 hp PZL-130 Orlik made its public début at Paris in June 1985, preliminary details were released of its turboprop derivative, which was to be developed by Airtech Canada in Ontario. The second flying prototype of the PZL-130 was taken to Canada for conversion and flight testing. Its maiden flight with the PT6A engine took place in August 1986, but unfortunately this aircraft crashed late in the following January, reportedly during a demonstration flight. At time of writing it was stated by PZL-Warsawa – Okecie that the programme would continue and that a replacement prototype would fly shortly. Manufacture will take place (in the event of a production order) in Poland, but Airtech Canada will be responsible for overseas marketing, and for final assembly, avionics installation, painting and flight testing of export PZL-130Ts.

Fuel capacity appears to be unchanged from that of the PZL-130, but the pneumatic system for undercarriage retraction has been replaced by hydraulics, and the aircraft has four hardpoints rather than two.

Powerplant: *P&WC PT6A-25A flat-rated at 550 shp.*
Dimensions: *wingspan 26 ft 3 in (8.0 m), length 28 ft 5 1/2 in (8.68 m), height 11 ft 7 in (3.53 m), wing area 132.4 sq ft (12.3 m²).*
Weights: *empty 2536 lb (1150 kg) internal fuel 730 lb (331 kg), normal take-off 3196 lb (1450 kg), max external load 1411 lb (640 kg), max take-off 4358 lb (1977 kg).*
Performance: *max level speed at sea level 236 knots (438 km/hr), at 15,000 ft (4570 m) 256 knots (475 km/hr), max rate of climb 3130 ft/min (15.9 m/sec), servicing ceiling 33,000 ft (10,000 m), range with external tanks 1198 nm (2220 km).*

The Turbo-Orlik is shown here with two 33 imp gal (150 litre) external tanks. (Pezetel)

EMBRAER EMB-312 Tucano

The Tucano was the first military trainer to be designed from the outset for a turboprop engine. It was also the first to use an engine as powerful as 750 shp, and the first to be fitted as standard with ejection seats. Design work began in 1978, and the first prototype had its maiden flight on 16 August 1980.

This aircraft is unusual in having a single-piece hood with integral windscreen. It is equipped with Martin-Baker BR8LC lightweight ejection seats, and has provisions for four wing pylons.

The Brazilian Air Force order was for 118 aircraft, designated T-27, with an option on a further 50, which may be powered with the more

Powerplant: P&WC PT6A-25C rated at 750 shp.
Dimensions: wingspan 36 ft 6$^{1}/_{2}$ in (11.14 m), length 32 ft 4$^{1}/_{2}$ in (9.86 m), height 11 ft 1$^{1}/_{2}$ in (3.40 m), wing area 208.82 sq ft (19.4 m^2).
Weights: basic empty 3990 lb (1810 kg), internal fuel 1166 lb (529 kg), max external load 2205 lb (1000 kg), clean take-off 5622 lb (2550 kg), max take-off 7000 lb (3175 kg).
Performance: max level speed at sea level 230 knots (426 km/hr), at 8000 ft (2440 m) 242 knots (448 km/hr), max cruising speed at 10,000 ft (3000 m) 222 knots (411 km/hr). max rate of climb 2230 ft/min (11.33 m/sec), service ceiling 24,800 ft (7560 m), range on internal fuel 1110 nm (2055 km), with two external tanks 1855 nm (3435 km).

A Tucano of the Venezuelan Air Force. (Embraer)

powerful Garrett engine of the Shorts version, for use in close support duties. The Garrett-engined EMB-312G1 first flew on 28 July 1986. Another assembly line has been established by the Arab Organisation for Industrialisation (AOI) at Heliopolis in Egypt, to produce initially 40 Tucanos for Egypt and 80 for Iraq, with options on a further 40 and 20 aircraft respectively. The first export order was for Honduras (which now has 12), and the Egyptian contract was followed by an order for 30 for Venezuela, then 20 for Peru and 30 for Argentina. The Shorts order for 130 aircraft brings the apparent total of firm orders and options to 570, but Embraer quotes a total of 586 aircraft.

IAR-825TP Triumf

The IAR-825TP Triumf. (Roy Braybrook)

The IAR-825TP may be regarded as a tandem-seat trainer derivative of the IAR-823 two/five-seat light aircraft, which is currently used by the Romanian Air Force for primary/basic flying training. It appears, however, that only the wings and undercarriage of the -823 are retained, most of the airframe being new.

The IAR-825TP made its first flight on 12 June 1982 and appeared at Farnborough three months later, preceding the piston-engined IAR-831,

Powerplant: P&WC PT6A-25C rated at 750 shp.
Dimensions: wingspan 32 ft 9¹/₂ in (10.0 m), length 29 ft 6 in (8.987 m), height 9 ft 11 in (3.025 m), wing area 161.5 sq ft (15.0 m²).
Weights: empty equipped 2756 lb (1250 kg), max take-off (aerobatic) 3748 lb (1700 kg), (utility) 5070 lb (2300 kg), (normal) 5842 lb (2650 kg), max external load 890 lb (400 kg).
Performance: max level speed at 13,100 ft (4000 m) 254 knots (470 km/hr), max cruising speed at 13,100 ft (4000 m) 237 knots (440 km/hr), max rate of climb 3150 ft/min (16.0 m/sec), service ceiling 29,500 ft (9000 m), range with 30 min reserves 755 nm (1400 km).

which was first shown at Paris in 1983. It was flown initially with a 680 shp agricultural engine, the PT6A-15AG, but the 750 shp PT6A-25C was planned for the production version. The IAR-825TP thus has the same engine as the Embraer Tucano, but it is a much smaller, lighter aircraft. Provisions are made for six underwing pylons and tiptanks.

Pilatus PC-9

The PC-9 represents a bold attempt by Pilatus to move forward from the 550 shp PC-7, leapfrog the 750 shp Embraer Tucano, and create a new category of turboprop trainer around an engine in the 1000 shp class. If successful, this new category would compete with the basic jet trainers,

The first prototype PC-9. (Pilatus)

performing their role at substantially reduced cost. The company might reasonably have expected to enjoy a monopoly in this field for several years, but Britain's AST.412, written around an improved Tucano, encouraged other manufacturers to compete with Pilatus, and the RAF order was eventually won by an improved Tucano promoted by Shorts.

The design of the PC-9 began in 1982, two years after the first flight of the Tucano, and the new Swiss aircraft had its maiden flight on 7 May 1984. Although the PC-7 might be said to have provided a starting-point for the new design, the PC-9 is a larger, heavier aircraft that has only 10 per cent commonality with its predecessor. The PC-9 features a stepped tandem-seat cockpit with Martin-Baker CH 11A ejection seats, a modified wing with new ailerons, an airbrake, doors to cover the retracted main undercarriage, and larger wheels with high pressure tyres.

Orders to date include four for Burma, 30 for Saudi Arabia, 67 for Australia (mostly to be built locally) and a total of around 25 for two undisclosed customers. Some reports suggest that this quantity refers to 4 for Angola and '15 +' for Iraq, although the latter has already received 52 PC-7s and has orders and options for 100 Egyptian-built Tucanos.

Powerplant: 1150 shp P&WC PT6A-62 derated to 950 shp.
Dimensions: wingspan 33 ft 2¹/₂ in (10.12 m), length 32 ft 11¹/₂ in (10.05 m), height 10 ft 8¹/₂ in (3.26 m), wing area 175.3 sq ft (16.29 m²).
Weights: empty equipped 3715 lb (1685 kg), max take-off (aerobatic) 4960 lb (2250 kg), (utility) 7055 lb (3200 kg).
Performance: max level speed at sea level 268 knots (496 km/hr), at 20,000 ft (6100 m) 300 knots (556 km/hr), max rate of climb 4000 ft/min (20.33 m/sec), max operating altitude 25,000 ft (7600 m), ceiling 38,000 ft (11,600 m), range with 5 per cent reserves and 20 min loiter 830 nm (1540 km).

Shorts Tucano

The Shorts version of the Tucano is a derivative of the Embraer aircraft, with a more powerful Garrett engine, a stronger and more fatigue-resistant airframe, an airbrake, a modified cockpit, and different armament options. It was developed under an agreement signed by the two companies in May 1984, specifically to produce an aircraft to meet RAF needs. Selection of the Shorts Tucano was formally announced in March 1985, the contract covering 130 Tucano T Mk1s to be delivered from late 1986 to 1992.

In meeting RAF demands, the design load factor has been increased from 6G to 7G, and the safe fatigue life increased from 8000 to 12,000 flight hours. To meet British birdstrike requirements, a windscreen was to be introduced between the two cockpits, and the design impact speed increased from 210 to 268 knots (390 to 497 km/hr). The avionics fit has

Powerplant: 1100 shp Garrett TPE331-12B.
Dimensions: wingspan 37 ft 0 in (11.28 m), length 32 ft 4 in (9.86 m), height 11 ft 2 in (3.40 m), wing area 208 sq ft (19.33 m²).
Weights: basic empty 4447 lb (2017 kg), internal fuel 1223 lb (555 kg), max take-off weight (aerobatic) 5842 lb (2650 kg), (utility) 7220 lb (3275 kg).
Performance: max level speed at 10,000 ft (3000 m) 274 knots (507 km/hr), economical cruising speed at 20,000 ft (6100 m) 220 knots (407 km/hr), max rate of climb 3510 ft/min (17.84 m/sec), service ceiling 34,000 ft (10,365 m), range on internal fuel 940 nm (1740 km), on external tanks 1800 nm (3335 km).

The Shorts Tucano in RAF colours. (Shorts)

been modified to suit RAF needs.

The first flight of a Garrett-engined Tucano took place in Brazil on 14 February 1986, and this aircraft flew for the first time in the UK on 11 April 1986. The first production Tucano assembled by Shorts was reportedly flown on 30 December 1986, though the formal roll-out took place on 20 January 1987. The programme initially ran somewhat behind schedule, but it was anticipated that the delays would be eliminated by the time the 30th aircraft was due for delivery in April 1988. Aside from a possible follow-on order from the RAF, Shorts anticipate selling 200 in the export market.

Chapter 7

Jet Trainers

AN EARLIER SECTION of this book has discussed one of the most recent developments in the jet trainer field, namely the advent of a very small and relatively low-priced aircraft (the Promavia Jet Squalus) to compete directly with the products at the top end of the turboprop range (the PC-9 and Shorts Tucano).

This is, however, only one of a series of developments and events that have served to shape the jet trainer spectrum. For example, the Jet Squalus may be regarded as the latest attempt to produce a true replacement for the highly successful Fouga CM-170 Magister, of which a total of 929 were built in France, Germany, Israel, and Finland in the period 1953-1969 for service in 16 countries.

The Fouga 90 was Aérospatiale's attempt to replace the Magister with a derivative design. The wing was retained virtually unchanged, but the engines were completely new, the fuselage was redesigned to allow the installation of ejection seats and a raised rear cockpit, and all the systems were modernised. The two engines were 1,545 lb (700 kg) Turboméca Astafan high-bypass turbofans, in which thrust was varied by adjusting the pitch of the fan blades. The Fouga 90 first flew on 20 August 1978, and gave a maximum sea level speed of 345 knots (640 km/hr). The French Air Force rejected this concept, however, and issued requirements that led to the piston-engined Aérospatiale Epsilon.

What happened in the UK in replacing the Jet Provost might be regarded as a re-run of the Magister replacement affair. Although it was common knowledge that the JP would run out of airframe life in the late 1980s, the RAF delayed issuing an AST (air staff target) for a replacement

The Fouga 90, a proposed
Magister replacement.
(Aérospatiale)

A Jet Provost T Mk5. (RAF
Support Command)

until June 1983, effectively leaving British Aerospace to develop a replacement in advance of the AST. The Warton design team first projected the Eaglet, which was a Strikemaster derivative with a JT15D or TFE731 turbofan in place of the original Viper turbojet.

Responsibility for the new basic trainer was then switched to Brough, where a completely new design, designated P.164, was proposed, using a JT15D similar to that used in the S.211. The project began life with side-by-side seating, but later drafts of the AST came down in favour of a tandem arrangement, so the aircraft was modified accordingly. Since Whitehall favoured a turboprop, such an aircraft was designed under the designation P.164-109, the turbofan proposal being the P.164-108.

In view of the rapidly shortening timescale, the MoD then demanded an off-the-shelf aircraft, and in March 1984 BAe abandoned its own projects in favour of supporting the Pilatus PC-9. The instructors at the flying training establishments reportedly favoured a turbofan, but cost considerations appear to have prevailed, and the turboprop Shorts Tucano was selected.

The propeller may have won over the turbofan in France and Britain, but the USAF currently remains intent on jet training. Planning for an

The P.164-108, British Aerospace's proposal for a Jet Provost replacement. (BAe, Brough)

aircraft to replace the Cessna T-37B began in the late 1970s. The T-37 (which first flew on 12 October 1954) had been a useful aircraft, but its lack of pressurisation restricted it to low altitude, it had only a limited capability in bad weather, and its turbojets restricted range and endurance. It was also painfully noisy for ground crew, being known as 'the world's largest dog whistle'. It had originally been designed for a life of 8,000 hours, and at the time of the Next Generation Trainer (NGT) studies it was assumed that it could not be taken beyond 15,000.

In specifying the NGT, the USAF wanted to retain twin engines and side-by-side seating, but it also wanted quiet, economical turbofans, full anti-icing provisions, pressurisation, and a cruise speed of 300 knots (556 km/hr) at 25,000 ft (7600 m). The RFP (request for proposals) was issued in 1981. While the preliminary planning had been taking place, President Carter had been buying no new aircraft, and it was widely assumed that the NGT would be a turbofan T-37, sometimes referred to as the TFT-37D. This would have had a gross weight of around 6400 lb (2900 kg). On the other hand, Rockwell was meantime promoting its Nova (near-term optimum value aircraft) project, which in early 1981 was expected to gross only 5325 lb (2115 kg) and would have been correspondingly less expensive to operate.

The prospect of substantial savings on life-cycle costs opened out the contest, and several other companies became interested. Vought proposed the Eaglet, based on Germany's ducted fan experience with the Fantrainer, and using as powerplant a pair of Allison 250-C30s or P&WC PT6B-34s, claimed very low fuel consumption and no engine development costs.

Gulfstream American (now Gulfstream Aerospace, a subsidiary of Chrysler) proposed a derivative of its Peregrine, which had first flown on 22 May 1981. The NGT version was to have two 1210 lb (550 kg) Williams FJ44 turbofans in place of the original single JT15D. Rockwell's Nova was also designed around a pair of Williams engines, although various alternatives were offered. General Dynamics proposed the relatively conventional Model 210 with a shoulder wing and a swept vertical tail.

On 2 July 1982 Fairchild Republic was named the winner, with a twin-tail design powered by a pair of Garrett TFE76 turbofans, later designated F109s. The initial contract covered two prototype T-46As and options on a further 54 aircraft out of an anticipated production total of 650. It may be noted that in replacing the T-37, its 75-hour syllabus would probably have been expanded to around 100 hours, to decrease the utilisation of the T-38, on which students currently spend 100 hours.

The first of two prototype T-46As had its maiden flight on 15 October 1985. It had an empty weight of 5587 lb (2534 kg) and a take-off weight of 7310 lb (3315 kg). It could reach 390 knots (723 km/hr) at 29,500 ft (9000 m), and had a service ceiling of 45,400 ft (13,840 m).

Unfortunately, the T-46A ran into development problems, and it fell behind schedule. When the USAF was forced to economise, the service declined to take up the options on production T-46As beyond the first

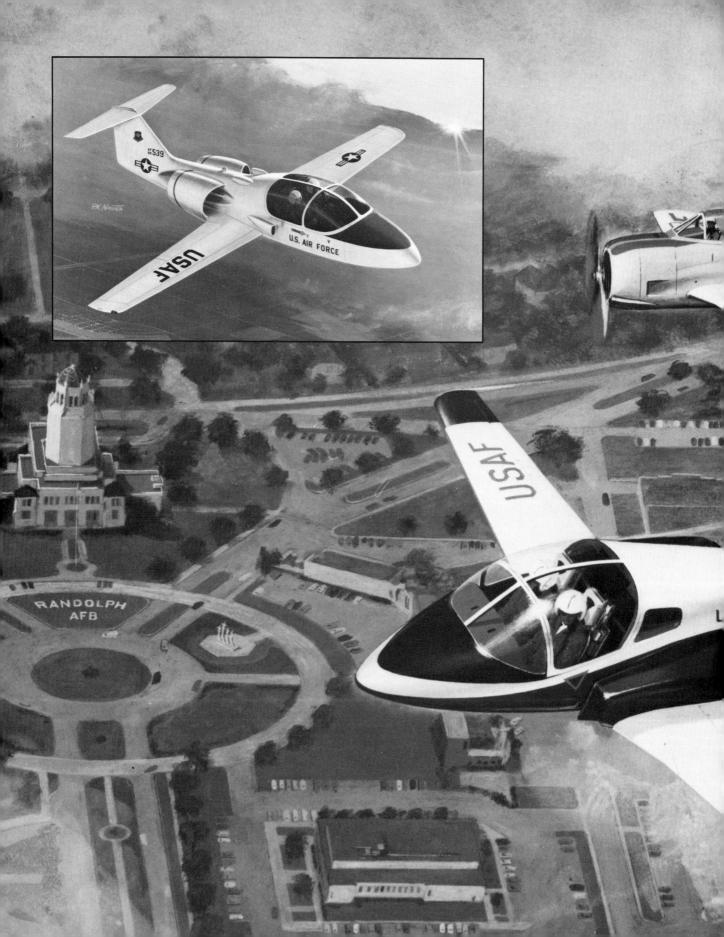

Above: The Gulfstream Peregrine was flown as a private venture, but failed to win support. (Gulfstream Aerospace)

Inset left: Vought's Eaglet NGT proposal was based on the Fantrainer. (Vought)

Left: A Rockwell artist's impression of the Nova project over Randolph AFB with (front to rear) a T-28, T-6, and BT-9, all earlier North American trainers. (Rockwell International, North American Aircraft Div)

batch of 10. This decision was made easier by new estimates for the life of the T-37 airframe, indicating that it was safe to 18,000 hours, and that a service life extension programme (SLEP) could take it to 33,000.

The T-46A programme was formally terminated on 13 March 1987 after negotiations on the cancellation of the 10 Lot 1 aircraft. The USAF will now spend $300,000 per aircraft on the SLEP for the T-37B, and is considering retrofit of the F109 engine.

In October 1986 Congress ordered that a new trainer competition should take place, with a flyoff in 1987 between the T-46A, an existing T-37, an upgraded T-37, and any other aircraft capable of meeting the NGT requirement. The new technology T-37 or 'NTT-37' would use the F109 engines of the T-46A, and would have new avionics and instruments, a pressurised cockpit, and a new vertical tail to improve control in crosswinds.

At time of writing the USAF has asked to be relieved of the demand to carry out a new contest, and there is speculation that the new (Democrat-led) Congress will review the whole affair, looking more favourably on the T-46A. In the meanwhile, the USAF has made it clear that it owns all the data and tooling for the T-46A, and may wish to put manufacture out to tender at some future date.

Other aspects of USAF pilot training appear to be progressing more smoothly. In 1980, going against the tradition for a unified training scheme that produced a 'universally assignable' pilot, the USAF in principle approved the ideas of 'dual-track' or specialised undergraduate pilot training (SUPT), in which the fighter, attack and reconnaissance (FAR) pilots would have a different syllabus from the tanker, transport and bomber (TTB) pilots. Initially the only difference was in the final month of T-38 training, the FAR students learning to fly four-ship formations while the TTB students did more instrument flying.

In terms of a more major change, the dual-track concept is part of the

The second development, test and evaluation Fairchild T-46A, photographed during its maiden flight on 29 July 1986. (TSgt Mike Hinson, USAF).

Top right: A big question mark hangs over the eventual replacement of the Northrop T-38 Talon, one of the great classic advanced trainers, shown here as it equipped the USAF Thunderbirds aerobatic team. (Northrop).

Right: An MDC artist's impression of what the T-45 might look like when modified for USAF use. (McDonnell Douglas)

five-year defence plan for Fiscal Year (FY) 1989, with initial operational capability (IOC) in 1991. To provide this capability, the USAF plans to acquire (beginning in FY89) some 215 TTB training aircraft in the form of off-the-shelf business jets. The introduction of these aircraft is expected to reduce annual utilisation of the T-38 by around 60 per cent.

It is also relevant to note that one of the options open to the USAF in replacing the T-37 is the US Navy's T-45 Goshawk, the McDonnell Douglas variant of the British Aerospace Hawk. This aircraft would be more expensive than the T-46, but it would be able to carry out a much larger part of the training syllabus. The T-45 might, in fact, replace both the T-37 and T-38 in USAF service in the same way that it has replaced both the T-2 and TA-4 in US Navy use.

Caproni Vizzola C22J Ventura

The Italian firm of Caproni is one of the world's oldest manufacturers of aircraft, having been founded in 1910. Since 1983 it has been part of the Agusta group.

The first pre-series C22J Ventura, which differs from the prototype in having tiptanks on reduced-span wings, and the more powerful TRS18-1 engines. (Caproni)

Powerplant: two Microturbo TRS18-1 turbojets rated at 325 lb (148 kg) each. ***Dimensions:*** wingspan 30 ft 2^1/$_2$ in (9.20 m), length 20 ft 6^1/$_2$ in (6.264 m), height 6 ft 2 in (1.88 m), wing area 82.35 sq ft (7.65 m²).
Weights: empty 1628 lb (739 kg), max external load 550 lb (250 kg), max take-off 2760 lb (1255 kg).
Performance: max level speed at sea level 280 knots (518 km/hr), economical cruising speed at 10,000 ft (3000 m) 175 knots (324 km/hr), max rate of climb 1930 ft/min (9.8 m/sec), max operating altitude (FAR limited) 25,000 ft (7620 m), service ceiling 29,500 ft (9000 m), max range 650 nm (1200 km).

The C22J is a trainer derivative of the A-21SJ Calif, a two-seat sailplane with a single turbojet engine, which first flew in May 1977. Although this powered sailplane provided a starting point for the new design, the C22J airframe is largely new. The prototype, fitted with two 220 lb (100 kg) Microturbo TRS18-046 turbojets, first flew on 21 July 1980. Subsequent aircraft have uprated engines, tiptanks, and provisions for two underwing pylons, with a second pair available as an option.

The original wooden prototype Microjet 200. (Creuzet)

Powerplant: two Microturbo TRS18-1 turbojets rated at 325 lb (148 kg).
Dimensions: wingspan 24 ft 9½ in (7.56 m), length 21 ft 10½ in (6.665 m), height 7 ft 5½ in (2.27 m), wing area 65.87 sq ft (6.12 m²).
Weights: empty 1698 lb (770 kg), internal fuel 750 lb (340 kg), max take-off (aerobatic) 2513 lb (1140 kg), (utility) 2866 lb (1300 kg).
Performance: max level speed and max cruising speed at 18,000 ft (5500 m) 250 knots (463 km/hr), economical cruising speed 210 knots (389 km/hr), max rate of climb 1705 ft/min (8.66 m/sec), service ceiling 30,000 ft (9150 m), range with 20 min reserves 470 nm (870 km).

Microjet 200B

The microjet is an unusual design, combining side-by-side seating with the instructor set back 21.65 inches (55 cm), a vee-tail, and two small engines fed by NACA-type flush inlets in the fuselage sides. The design and early development were carried out by Microturbo to demonstrate the usefulness of the company's TRS18 series of turbojets.

The prototype Microjet 200 was of wooden construction, and first flew on 24 June 1980, powered by TRS18-048s of 225 lb (102 kg) each. Several changes then took place, as the more powerful TRS18-1 became available, and wooden construction was replaced by the use of composites for the wings and tail, and metal for the fuselage. The resulting Microjet 200B first flew on 19 May 1983, and the second pre-series aircraft on 5 January 1985, with two underwing hardpoints.

Further development of the project has been taken over by Marmande Aèronautique, part of the Creuzet group. It is envisaged that the TRS18 will be progressively uprated to 360 lb (163 kg) and later 450 lb (204 kg).

Promavia Jet Squalus

The president of Belgium's Promavia SA is André Delhamende, whose Aspair company acquired export rights to the SF.260 and was responsible for the sales of more than 600 of these aircraft. In 1982 he formed Promavia and organised a study of the potential market for military trainers. This study identified a substantial demand for a low-cost basic jet trainer with side-by-side seating and the latest engine technology. Promavia therefore commissioned Italy's General Avia (under the direction of Dr Ing Stelio Frati) to design a lightweight training aircraft around the new Garrett F109 turbofan, and to construct two prototypes for use in certification and demonstration.

The prototype of the Jet Squalus, which was shown statically at Farnborough in 1986. (Promavia)

> *Powerplant:* Garrett TFE109-1 turbofan rated at 1330 lb (603 kg).
> *Dimensions:* wingspan 29 ft 8 in (9.04 m), length 30 ft 8$\frac{1}{2}$ in (9.36 m), height 11 ft 9$\frac{1}{2}$ in (3.60 m), wing area 146.06 sq ft (13.58 m^2).
> *Weights:* empty 2865 lb (1300 kg), max take-off (aerobatic) 4410 lb (2000 kg), (normal) 5290 lb (2400 kg).
> *Performance:* max level speed at 14,000 ft (4270 m) 315 knots (584 km/hr), max rate of climb 3000 ft/min (15.24 m/sec), ceiling 37,000 ft (11,300, m), max operating ceiling 25,000 ft (7600 m), range on internal fuel 1000 nm (1850 km).

The first prototype was completed in August 1986 and appeared statically at Farnborough in the following month. It first flew on 30 April 1987. This aircraft is not pressurised, but it is understood that the second will have pressurisation, four hardpoints, and possible uprated engines. The existing F109 has a thrust of 1330 lb (603 kg), but Garrett is considering uprating it to 1600 lb (725 kg) and later 1800 lb (815 kg). The latter rating would provide an initial climb rate of 4000 ft/min (20.33 m/sec). The same thrust could be provided by the Williams FJ44.

If Promavia secures orders for production in Belgium, the Jet Squalus will be manufactured by Sonaca. On the other hand, it is also being promoted in the US as a possible T-46 substitute by Rockwell, where it would be built if ordered by the USAF.

SIAI-Marchetti S.211

The S.211 might be regarded as Siai's attempt to develop a basic jet trainer that can repeat the remarkable success of the SF.260 in the international

Internal layout of the S.211. (SIAI-Marchetti)

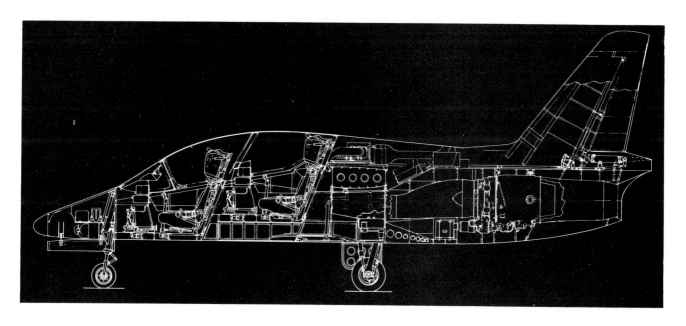

The S.211 in flight, illustrating the raised position of the instructor in the rear cockpit. (SIAI-Marchetti)

Powerplant: P&WC JT15D-4C turbofan rated at 2500 lb (1135 kg).
Dimensions: wingspan 27 ft 7 in (8.43 m), length 31 ft 1 in (9.50 m), height 12 ft 6 in (3.80 m), wing area 135.6 sq ft (12.60 m²).
Weights: empty equipped 3747 lb (1700 kg), internal fuel 1370 lb (622 kg), max take-off clean 5952 lb (2700 kg), with external stores 6835 lb (3100 kg).
Performance: max cruising speed at 20,000 ft (6000 m) 360 knots (667 km/hr), max rate of climb 4200 ft/min (21.4 m/sec), service ceiling 40,000 ft (12,200 m), max range on internal fuel 900 nm (1670 km), ferry range with external tanks 1100 nm (2040 km).

market. The design aim (from the start of work in 1975) was to combine low unit cost with advanced technology, using the P&WC JT15D turbofan and a supercritical wing section. The first of two prototypes flew on 10 April 1981, and series production was launched on the basis of an order from Singapore for 10 aircraft with an option on a further 20. Smaller orders from Haiti and one other Latin American country took the total to around 40 units.

Aside from hydraulically-boosted ailerons, the flying controls are manually operated. The flaps are activated electrically, and the airbrake hydraulically. The wing has four hardpoints.

Aero L-39 Albatross

Czechoslovakia's L-39 is a straight-wing basic trainer powered by the Ivchenko AI-25 turbofan used in the Yak-40 feederliner transport. It first flew on 4 November 1968, and entered service in the spring of 1974, having been selected as the standard jet trainer for most of the Warsaw Pact air forces. It has subsequently been exported to many other countries, including Afghanistan, Cuba, Ethiopia, Iraq, Libya, Syria and Vietnam. It has thus proved itself over a wide range of conditions, from extreme cold in Russia to very high temperatures in Iraq. Reports indicate that more than 1500 have been built, and that production continues at around 100 aircraft per year.

The L-39 is unusual in having been designed to operate from grass fields, which explains the location of the intakes above the wing, to minimise debris ingestion. The flying controls are manually operated, although the aircraft has a hydraulic system for the flaps, undercarriage, airbrake, etc. It has been built in various models from the original L-39, which had no armament provisions, to the L-39Z, which has four pylons and a 23 mm twin-barrel GSh-23 cannon with 150 rounds. There have been unconfirmed reports of an uprated engine producing 5300 lb (2400 kg) of static thrust.

Powerplant: Ivchenko AI-25TL turbofan rated at 3792 lb (1720 kg).
Dimensions: wingspan 31 ft $0^1/2$ in (9.46 m), length 39 ft $9^1/2$ in (12.13 m), height 15 ft $7^1/2$ in (4.77 m), wing area 202.36 sq ft (18.8 m²).
Weights: empty weight (L-39) 7625 lb (3459 kg), (L-39Z) 8060 lb (3656 kg), internal fuel 1817 lb (824 kg), (with tiptanks 2160 lb (980 kg), max external load 2425 lb (1100 kg), max take-off (clean) 10,030 lb (4549 kg), with tiptanks) 10,365 lb (4700 kg), (with external stores) 12,450 lb (5646 kg).
Performance: max level speed at sea level (clean) 378 knots (700 km/hr), at 5000 ft (1500 m) 405 knots (750 km/hr), max rate of climb 4035 ft/min (20.50 m/sec), service ceiling 36,080 ft (11,000 m), ferry range with two underwing tanks and 5 per cent reserves 945 nm (1750 km).

An L-39Z with ventral 23 mm GSh-23 cannon, drop tanks and K-13 Atoll air-air missiles. (Aero Vodochody)

Cutaway drawing of the L-39Z.
(Aero Vodochody)

Above: Front cockpit of the L-39.
(Aero Vodochody)

CASA C-101 Aviojet

The development contract for the C-101 was signed in September 1975, the aim being to produce an aircraft to replace both the HA200/220 and the T-33 in Spanish Air Force service. The first of two prototypes had its maiden flight on 27 June 1977, and the C-101EB with the 3500 lb (1587 kg) Garrett TFE731-2-2J turbofan began to be delivered as the E-25 Mirlo (Blackbird) in March 1980.

The C-101 is a comparatively large aircraft with generous fuel capacity (eliminating the need for external tanks), but it was originally somewhat underpowered, hence it has been developed in a series of stages with progressively more thrust. The C-101BB is an armed export version with the 3700 lb (1680 kg) TFE731-3-1J engine, as sold to Chile and Honduras. The C-101CC is a light attack variant with a TFE731-5-1J giving up to

Powerplant: *Garrett TFE731-5-1J turbofan with a normal rating of 4300 lb (1950 kg) and a military power reserve (MPR) of 4700 lb (2130 kg).*
Dimensions: *wingspan 34 ft 9 in (10.60 m), length 41 ft 0 in (12.50 m), height 13 ft 11 in (4.25 m), wing area 215 sq ft (20 m²).*
Weights: *empty equipped 7715 lb (3500 kg), normal fuel 2972 lb (1348 kg), max internal fuel using outer wing tanks 4148 lb (1880 kg), max external load 4960 lb (2250 kg), training mission weight (partial fuel) 10,075 lb (4570 kg), max take-off weight 13,890 lb (6300 kg).*
Performance: *max level speed at sea level 415 knots (769 km/hr), at 15,000 ft (4757 m) 450 knots (834 km/hr), economical cruising speed at 30,000 ft (9150 m) 354 knots (656 km/hr), max rate of climb using MPR 6100 ft/min (31.0 m/sec), service ceiling 42,000 ft (12,800 m), ferry range with 30 min reserves 2000 nm (3700 km).*

The C-101DD with nose radar, 30 mm DEFA cannon and underwing stores. (CASA)

4700 lb (2130 kg), and was sold to Chile and Jordan. The C-101DD is a development of the 101CC, with a comprehensive nav-attack system.

Aside from its remarkable fuel capacity, the C-101 is noteworthy for its six pylons and its neat gunpack, which uses the space under the rear seat, and can accommodate either a single 30 mm DEFA cannon or two 12.7 mm machine guns.

The following specification relates to the C-101CC:

FMA IA-63 Pampa

The IA-63 was developed as a replacement for the MS.760 Paris in Argentine service. The programme began in 1979, and FMA (Fábrica Militar de Aviones) awarded Dornier a contract to assist with the design. The involvement of the German company explains why the aircraft that first flew on 6 October 1984 looked like a small, straight-wing, single-engine Alpha Jet. The IA-63 has provisions for five hardpoints, including a fuselage station for a 30 mm DEFA cannon pod.

The initial Argentine Air Force requirement is for 64 aircraft, but it is anticipated that a further 40 may be purchased. These later aircraft may be used specifically for close support duties, and be fitted with an uprated engine. There appear to be several export prospects in Latin America, including Brazil.

In late 1986 FMA was renamed 'Aérea Material Córdoba'.

Aermacchi MB.339

The MB.339 was designed primarily to replace the MB.326 basic jet trainer in Italian Air Force service, and to attempt to repeat the success of the MB.326 in marketing terms, some 784 of the earlier type having been produced for 11 customers.

The new design retained the 4000 lb (1815 kg) Viper 600 series, which had been fitted to the MB.326L and the single-seat MB.326K. In other respects the MB.339 is virtually a brand-new design, although aerodynamic development benefited from having the MB.326 as a starting point. Relative to the MB.326, the wing is strengthened, the rear cockpit is raised so that the instructor can use a gunsight, the cockpits and avionics are modernised, the fin is enlarged, and the flying controls and cabin conditioning are improved. The parts commonality between the two aircraft is negligible.

The first MB.339 flew on 12 August 1976, and deliveries of 100 aircraft for the Italian Air Force began in 1979. In the IAF this aircraft takes students from the SF.260 to the G.91T, which is used for advanced flying training. Exports currently total 52 aircraft for Argentina, Dubai, Malaysia, Nigeria and Peru.

The aircraft delivered to the Italian Air Force are designated MB.339A. The MB.339B has the uprated Viper 680-43 of 4450 lb (2018 kg). The MB.339C has a comprehensive nav-attack fit. The MB.339K Veltro II

Powerplant: *Garrett TFE731-2 rated at 3500 lb (1587 kg).*
Dimensions: *wingspan 31 ft 9 in (9.69 m), length 35 ft 9 in (10.90 m), height 14 ft 1 in (4.30 m), wing area 168 sq ft (15.63 m^2).*
Weights: *empty equipped 5791 lb (2627 kg), normal internal fuel 1745 lb (792 kg), max internal fuel 2646 lb (1118 kg), max external load with normal fuel 2557 lb (1160 kg), normal clean take-off 8377 lb (3700 kg), max take-off 11,023 lb (5000 kg).*
Performance: *max level speed at sea level 400 knots (740 km/hr), max cruising speed at 13,000 ft (4000 m) 403 knots (747 km/hr), max rate of climb 5300 ft/min (27.0 m/sec), service ceiling 42,300 ft (12,900 m), range 810 nm (1500 km).*

Prototype IA-63 Pampa. (FMA)

The MB.339 with two 30 mm DEFA gunpods, four bombs and cylindrical tiptanks. (Aermacchi)

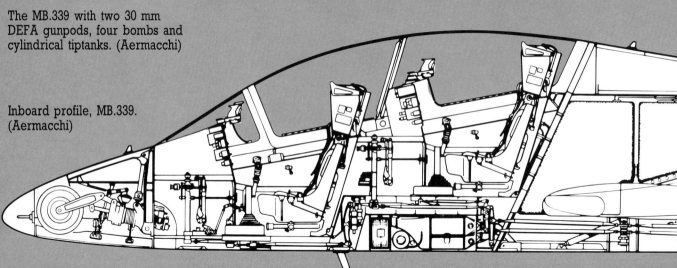

Inboard profile, MB.339. (Aermacchi)

Powerplant: *Rolls-Royce Viper 680-42 turbojet, rated at 4450 lb (2018 kg).*
Dimensions: *wingspan with tiptanks 36 ft 9½ in (11.22 m), length 36 ft 0 in (10.97 m), height 12 ft 9½ in (3.9 m), wing area 207.7 sq ft (19.3 m²).*
Weights: *empty 6962 lb (3158 kg), internal fuel 3060 lb (1388 kg), max external load 4000 lb (1815 kg), take-off (training mission) 10,410 lb (4720 kg), max take-off 14,000 lb (6350 kg).*
Performance: *max level speed at sea level 487 knots (902 km/hr), max rate of climb 8500 ft/min (43.2 m/sec), ferry range with two underwing tanks and 10 per cent reserves 1190 nm (2200 km).*

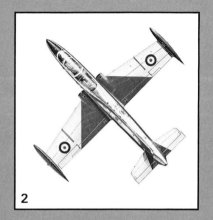

2

(Greyhound) is a single-seat derivative with two 30 mm DEFA cannon in the lower front fuselage. At time of writing the B, C, and K do not appear to have been ordered in series.

The specification lower left refers to the MB.339B:

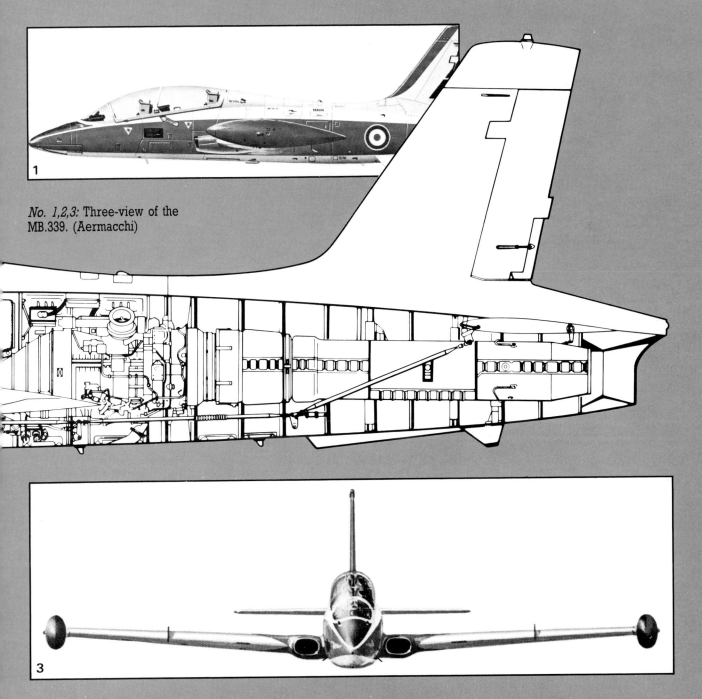

No. 1,2,3: Three-view of the MB.339. (Aermacchi)

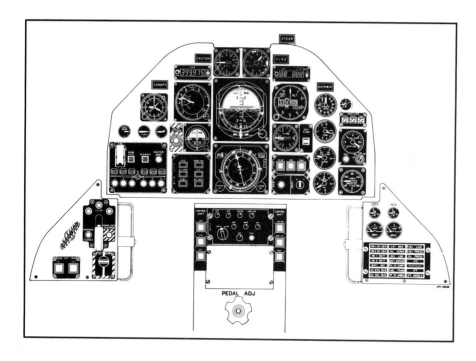

Instrument layout, MB.339.
(Aermacchi)

Soko G-4 Super Galeb

Yugoslavia's G-4 was designed to replace the Soko G-24 Galeb (Seagull) and Lockheed T-33 in that country's service. The basic philosophy was that trainers such as the Hawk and Alpha Jet are too expensive and sophisticated, and are not suitable for direct transition from a piston-engined primary trainer. The G-4 was therefore aimed at the upper end of the L-39/C-101/MB.339 category, allowing it to take students directly from the 180 hp UTVA-75 to a combat aircraft such as the MiG-21. In order to retain low cost while providing the highest possible performance, the G-4 was equipped with the most powerful version of the Rolls-Royce Viper turbojet (which Soko had used in earlier form in the G-2 series and the single-seat Jastreb) and a moderately swept wing. Quarter-chord sweep is 22 degrees.

The first of two prototypes had its maiden flight on 17 July 1978, and the aircraft was exhibited at Paris in 1983, at which stage it was entering service. Unusual features include a braking parachute as standard, low pressure tyres, and provisions for two assisted take-off rockets. To reduce cost, the aircraft is normally gravity-refuelled, although pressure refuelling is available as an option. The ailerons and tailplane are hydraulically powered. The rear seat is raised 9.85 in (25 cm), but this is not sufficient to allow the installation of a gunsight in the rear cockpit. The G-4 has provisions for four underwing pylons and a centre-line 23 mm GSh-23LY twin-barrel cannon pod with 200 rounds. Like the Viper engine, the GSh-23LY is made under licence. The "Y" indicating Yugoslavia. A single-seat version of the G-4 has been considered (corresponding to the

Jastreb variant of the G-2), but the intention currently appears to be to leave the close support role to the Orao or IAR-93 developed jointly with Romania.

Soko G-4 Super Galeb at Paris in 1983. (Roy Braybrook)

Powerplant: Rolls-Royce Viper 632 turbojet rated at 4000 lb (1815 kg).
Dimensions: wingspan 32 ft 5 in (9.85 m), length 38 ft 4 in (11.86 m), height 14 ft 0 in (4.28 m), wing area 210 sq ft (19.5 m²).
Weights: empty equipped 7165 lb (3250 kg), internal fuel 2990 lb (3250 kg), take-off (training) 10,495 lb (4760 kg), with 2975 lb (1350 kg) of external stores 13,472 lb (6110 kg), overload 13,958 lb (6330 kg).
Performance: max level speed at 20,000 ft (6000 m) 491 knots (910 km/hr), max rate of climb 5900 ft/min (30 m/sec), absolute ceiling 49,200 ft (15,000 m), range 437 nm (810 km).

Dassault-Breguet/Dornier Alpha Jet

The Alpha Jet was designed primarily to meet a joint Franco-German requirement for a training and close support aircraft. In the French Air Force it was to replace the T-33 in the advanced training role and the Mystère IVA in weapons training, while in the German Air Force it was to replace the G-91 in ground attack, reconnaissance and weapons training duties.

The joint requirement was announced in 1969, the Alpha Jet was selected in the following year, and the first prototype flew on 26 October 1973. It was not until late in 1975 however, that production was formally authorised, and the aircraft consequently did not reach full operational status until 1980. In spite of this delay, the Alpha Jet has sold reasonably well. Orders now total 500 for 10 countries, including 175 each for France and Germany.

The German Air Force version of the Alpha Jet. (Dassault-Breguet)

No. 1,2,3: Three-view of the Alpha Jet as operated by the French Air Force. (Dassault-Breguet)

1

There were originally two production versions: the pointed-nose German version with four pylons, full nav-attack fit and provisions for a ventral 27 mm Mauser cannon, and the rounded-nose French version with improved spinning characteristics, two pylons and provisions for a ventral 30 mm DEFA pod. Most export aircraft have the rounded nose, DEFA cannon, four pylons, and French avionics. More recently the 'new generation' NGEA aircraft has been produced for Egypt and the Cameroon, with a comprehensive nav-attack system and uprated Larzac 04-C20 engines of 3177 lb (1440 kg) in place of the standard 04-C6s of 2965 lb (1345 kg). As currently planned, the next stage of development is the Lancier with radar or other sensors in the nose, permitting day/night all-weather operation, and provisions for advanced weapons, such as the Aérospatiale AM.39 Exocet.

The specification opposite refers to the original standard:

Powerplant: *two SNECMA/ Turboméca Larzac 04-C6 turbofans of 2965 lb (1345 kg) each.*

Dimensions: *wingspan 29 ft 11 in (9.10 m), length (French a/c) 38 ft 10 in (11.85 m), (German a/c) 40 ft 3$\frac{1}{2}$ in (12.29 m), height 13 ft 9 in (4.19 m), wing area 188.4 sq ft (17.50 m^2).*

Weights: *empty equipped (trainer) 7375 lb (3345 kg), (attack) 7749 lb (3515 kg), internal fuel (training sortie) 3351 lb (1520 kg), max internal fuel 3593 lb (1630 kg), take-off clean (trainer) 11,023 lb (5,000 kg), max overload 16,000 lb (7250 kg).*

Performance: *max level speed at sea level 540 knots (1000 km/hr), max rate of climb 11,200 ft/min (57 m/sec), service ceiling 48,000 ft (14,630 m), ferry range with two underwing tanks 1400 nm (2600 km).*

2

3

British Aerospace Hawk

The Hawk was designed to fulfil RAF needs for a new training aircraft and to produce an affordable multi-role training and combat aircraft for the export market. This design was selected for development in 1971, and in the following year an order for 176 Hawk T1s was placed on behalf of the RAF. The first aircraft flew on 21 August 1974, just in time to make its début at Farnborough in the following month and deliveries to the service began in late 1976. The T1 is used both for advanced flying training (replacing the Gnat) and tactical weapons instruction (replacing the Hunter T7).

The Hawk might be regarded as a single engined, low wing Alpha Jet, but the British aircraft is somewhat less expensive, has a larger cockpit, and was designed for heavier loads and a longer fatigue life. Its development was facilitated by the extensive prior operating experience of its Adour engine in the Jaguar series.

The RAF's T1 is powered by a 5300 lb (2405 kg) Adour 151 and is equipped with two pylons and provisions for a ventral 30 mm Aden cannon. Some 88 T1s have been modified to T1A standard, with provisions

Line-up of Hawk T1s at RAF Valley. (Michael Stroud, BAe)

An RAF Hawk in war-role
configuration, with 30 mm cannon
and Sidewinder missiles. It is
painted air superiority grey.
(Geoffrey Lee, BAe)

for Sidewinder carriage in the low level airfield defence role.

The 50-series was similar to the T1, but had provisions for four pylons, two of which could take fuel tanks. The 60-series has the 5,700 lb (2585 kg) Adour 861. A total of 156 Hawks in the 50/60 series have been sold to eight nations, excluding the anticipated Swiss order. The 100-series is a proposal with a comprehensive nav-attack system, originally offered to

Venezuela. The 200-series is a single-seat derivative with a 5845 lb (2650 kg) Adour 871 and two 25 mm Aden 25s in the lower front fuselage. The first Hawk 200 flew on 19 May 1986, and the second on 24 April 1987.

The next major step in the development of the Hawk is the McDonnell Douglas T-45A Goshawk, a navalised derivative that in 1981 won the US Navy contest for an aircraft to replace the T-2C and TA-4J. The T-45A has a 5400 lb (2450 kg) Adour 861-49, which is designated F404-RR-400. Empty weight is increased to 9335 lb (4234 kg), and maximum gross is 12,700 lb (5760 kg). Full-scale development began in 1984, and first flight was planned for late 1987. The production of 307 aircraft is anticipated. The following specification relates to the Hawk 60-series:

An artist's impression of the T-45A Goshawk trainer for the US Navy. (McDonnell Douglas)

Powerplant: *Rolls-Royce/Turboméca Adour Mk 861 turbofan of 5700 lb (2585 kg).*
Dimensions: *wingspan 30 ft 10 in (9.39 m), length 38 ft 11 in (11.85 m), height 13 ft 2 in (4.0 m), wing area 179.64 sq ft (16.69 m²).*
Weights: *empty 7940 lb (3600 kg), internal fuel 2965 lb (1345 kg), max external load 6835 lb (3100 kg), take-off clean 11,280 lb (5115 kg), current take-off max 16,600 lb (7530 kg), planned max 18,960 lb (8600 kg).*
Performance: *max level speed at sea level: 560 knots (1038 km/hr), max rate of climb 11,800 ft/min (60 m/sec), service ceiling 48,000 ft (14,650 m), ferry range with two tanks: 2200 nm (4080 km).*

Abbreviations

AAC	Australian Aircraft Consortium
AFB	Air Force Base
AMRAAM	Advanced Medium-Range Air-Air Missile
AOI	Arab Organisation for Industrialisation
ASEAN	Association of South-East Asian Nations
AST	Air Staff Target
AVGAS	aviation gasoline
BAe	British Aerospace
CASA	Construcciones Aeronauticas SA
COIN	counter-insurgency
CTA	Companion Trainer Aircraft
D&P	development and production
ENAER	Empresa Nacional de AERonautica
FAR	Federal Aviation Regulations
FAR	fighter, attack and reconnaissance
FFA	Flug- und Fahrzeugwerke
FMA	Fábrica Militar de Aviones
FTS	Flying Training School
FUS	Flugzeug-Union-Sued
FY	fiscal year

GRP	glass reinforced plastic
HUD	head-up display
IOC	initial operational capability
JASDF	Japan Air Self-Defense Force
LDPG	low-drag, general purpose
MBB	Messerschmitt-Boelkow-Blohm
MoD	Ministry of Defence
MOGAS	motor-car gasoline
MPR	military power reserve
NACA	National Advisory Committee for Aeronautics (now NASA)
NATO	North Atlantic Treaty Organisation
NGEA	Nouvelle Génération Ecole/Appui
Nova	Near-term optimum-value Aircraft
PAC	Pacific Aerospace Corporation
P&WC	Pratt & Whitney Canada
R&D	research and development
RAAF	Royal Australian Air Force
RAF	Royal Air Force
RFB	Rhein-Flugzeugbau
R-R	Rolls-Royce
RTAF	Royal Thai Air Force
SUPT	Specialised Undergraduate Pilot Training
TTB	tanker, transport and bomber
UK	United Kingdom (of Great Britain and Northern Ireland)
USAF	United States Air Force
USN	United States Navy
VLCT	Very Low-Cost Trainer

Index